Walter C. Patterson was born in Canada in 1936. He took a postgraduate degree in nuclear physics at the University of Manitoba. In 1960 he moved to Britain, where he became involved in environmental issues. He joined the staff of Friends of the Earth in London in 1972, was their energy specialist until 1978 and was their lead witness at the Windscale Inquiry. Since 1978 he has been an independent commentator and consultant, dealing with energy and nuclear policy issues. He is energy consultant to Friends of the Earth, and was a witness at the Sizewell Inquiry. He is a regular contributor to a number of publications, including *New Scientist* and *The Guardian*, and he also appears on radio and television. He is an editorial adviser to the *Bulletin of the Atomic Scientists*. Walter C. Patterson is the author of *Nuclear Power* (1976, and since updated), *The Fissile Society* (1977), *Fluidized Bed Energy Technology: Coming to a Boil* (1978) and *The Plutonium Business and the Spread of the Bomb* (1984), which is also published by Paladin.

WALTER C. PATTERSON

Going Critical

PALADIN
GRAFTON BOOKS
A Division of the Collins Publishing Group

LONDON GLASGOW
TORONTO SYDNEY AUCKLAND

Paladin
Grafton Books
A Division of the Collins Publishing Group
8 Grafton Street, London W1X 3LA

Published by Paladin Books 1985

ISBN 0-586-08516-5

Printed and bound in Great Britain by
Collins, Glasgow

Set in Ehrhardt

With Love to Cleone,
who also likes the
Marx Brothers

Contents

'going critical': reaching the point at which a chain reaction becomes self-sustaining; starting up a nuclear reactor; asking embarrassing questions

Prologue: To err is nuclear . . .

Nobody's perfect. We all make mistakes. Sometimes we escape unscathed; more often we pay for our mistakes, in money or embarrassment or worse. Very few of us can get away with making a career out of making mistakes. One fortunate group in British society has, however, done so for upwards of three decades. Britain's nuclear power policy-makers have a track record of misjudgement, mismanagement and misfortune that would long since have brought any other enterprise to its knees, and made it a national laughing-stock. Yet these same policy-makers continue to exercise extraordinary influence in the uppermost echelons of government in Britain. Why should this be so?

One simple reason may be that the British public, and their elected representatives, remain somehow unaware of the débâcle that is British nuclear power. Even in 1985 official pronouncements from the nuclear industry and its government mentors almost invariably include some tribute to the alleged accomplishments of Britain's nuclear power establishment, its purported leading role in global nuclear technology. The truth, alas, is otherwise.

PART I

How Not to Build Nuclear Stations

1 The power struggle

'We have made the greatest breakthrough of all time.' Thus, twenty years ago, did Fred Lee, Minister of Power, herald the advent of Britain's second nuclear power programme. On 25 May 1965, announcing the government's latest nuclear plan, he told the House of Commons, 'We have hit the jackpot this time.' The jackpot in question was the advanced gas-cooled reactor, known for short as the AGR. Lee's statement gave the official go-ahead for the Central Electricity Generating Board to order the first of a new series of nuclear power stations. It also decreed that this new series was to be based on the advanced gas-cooled reactor, developed by the United Kingdom Atomic Energy Authority. The first station of this second series would be built on the south Kent coast. It would be known as Dungeness B.

Lee's ebullience was spectacularly misconceived. Dungeness B proved to be not so much a breakthrough as a breakdown. Ordered in August 1965, it did not even start up until December 1982 – more than seventeen years later. By that time its cost had reached more than five times the 1965 estimate, and its intended output had been scaled down more than 20 per cent. In 1985, two decades after the original order, the second reactor at the station had only just started up. Atomic Power Constructions, the company that won the Dungeness B contract in 1965, had by 1970 collapsed in total disarray, technical, managerial and financial.

In the lexicon of the British nuclear establishment Dungeness B has always been discounted as a desperately unfortunate anomaly. History, however, suggests the contrary. Dungeness B was not an anomaly. In the history of British nuclear power Dungeness B was merely the most conspicuous and long-running cock-up in a virtually endless catalogue of cock-ups, in planning, construction and operation. There were of course a few bright spots; but there was, and still is, a quite extraordinary variety of egregious embarrassments, cruelly at variance with the lofty long-range aspirations of those who first introduced nuclear power to Britain.

The military and political impact of nuclear fission caught British society, as it caught the whole world, unprepared. In the eyes of the politicians, nuclear research and nuclear materials were clearly too politically delicate to entrust to the general public and existing social institutions, like private industry. The British government, like other governments around the world, had to make things up as it went along. Even while British scientists and engineers were working overtime to produce Britain's own nuclear weapons, the idea of nuclear power had begun to percolate. But it had to concede precedence to the perceived urgency of the military demand for an atom bomb, indeed for a good many atom bombs.

From 1946 until 1954 British government responsibility for nuclear activities was in the hands of the so-called Ministry of Supply, a carry-over from World War II. The Ministry's Division of Atomic Energy, in several bureaucratic metamorphoses, oversaw the construction of the Atomic Energy Research Establishment at Harwell, the uranium factories at Springfields and the sprawling complex at Windscale, including two plutonium-production reactors and a chemical plant for separating the plutonium for bombs. It also set in train the construction of Calder Hall, a dual-purpose installation producing weapons-plutonium while generating electricity as a by-product. But influential advisers, notably Churchill's favourite scientist, Lord Cherwell, began lobbying vigorously for the creation of a new entity to run Britain's nuclear affairs. An official committee chaired by Lord Waverley made recommendations about a suitable structure for the new entity which were embodied in the Atomic Energy Act of June 1954. On 19 July 1954 the United Kingdom Atomic Energy Authority (UKAEA) came into being.

The UKAEA was unlike any other agency in Britain. Its financial and administrative powers were substantial; its control by Parliament was limited and tenuous. It was financed by a direct 'vote' of public funds, under conditions that offered little opportunity for MPs to find out what would be done with the money voted, either before or after it was spent. The very first estimate of the annual budget of the AEA put the sum likely to be required from Parliament at £53 million, at 1954 prices – a staggering sum to be found within an economy still trying to right itself after a devastating war.

At the time of its creation the primary responsibility of the AEA was the design, production and testing of nuclear weapons. Nevertheless, within weeks of its creation, the AEA was closeted with government officials, drawing up plans for Britain's first nuclear power stations. The result was a White Paper published in February 1955, entitled *A Programme of Nuclear Power*. In 1955 most of Britain's electricity supply – that in England and Wales – was provided by the Central Electricity Authority (CEA), soon to be transformed into the Central Electricity Generating Board (CEGB). Nuclear power stations would have to be integrated into the system operated by the electricity suppliers; accordingly, it was reasonable to expect that the electricity supply organization would be involved in the preliminary planning of the proposed nuclear power programme. It was not. On the contrary, the CEA was given just one month to comment on the draft White Paper before it was presented to Parliament. The Authority's off-hand attitude to its most important client was an early hint of what was to come.

As the title of the White Paper indicated, it proposed not just a single civil nuclear power station but a 'programme' of them: 1500–2000 megawatts of nuclear power stations, to be ordered by 1965. They would be based on the design used for the Calder Hall military station; they would use fuel in the form of rods of uranium metal, clad in a special magnesium alloy called 'Magnox'; and they would use carbon dioxide gas as 'coolant', to collect the heat from the chain reaction and carry it to the boilers. Stations of this design came to be known as Magnox stations. The stations would be ordered from and constructed by new industrial groups called 'consortia'. Four consortia, each led by one of the country's major manufacturers of heavy electrical plant – Associated Electrical Industries (AEI), the General Electric Company (GEC), English Electric and C. A. Parsons – had been set up by 1955; another was added in 1956. The Authority provided design information and held courses to train staff from the consortia in the subtleties of nuclear engineering. The idea behind having several consortia involved was to provide for competitive tendering for the new programme of nuclear plants. But no one appears to have asked whether or not the size of the programme could support so many ostensible competitors.

The first three civil Magnox stations were sited at Berkeley, in Gloucestershire, and at Bradwell, on the Essex coast, by the CEGB; and at Hunterston on Clydeside in Scotland, by the South of Scotland Electricity Board. The orders, placed before the end of 1956, went to the AEI consortium – known as the Nuclear Energy Company – for Berkeley; the Parsons consortium – known as the Nuclear Power Plant Company – for Bradwell; and the GEC consortium – known as the Atomic Energy Group – for Hunterston. In 1957 the fourth order, for a station at Hinkley Point in Somerset, went to the fourth consortium, led by English Electric, known as the Atomic Power Group. It would have been better if the four consortia had shown as little variety in their designs as they did in their names. Unfortunately, however, each consortium came up with its own design, differing significantly from that of its competitors. From that time onwards every new nuclear station ordered would be, in important respects, yet another prototype, with innovative design features differing from those of all its precursors. As it happened, these innovations did not by any means always signify improvements.

In October 1956 the British government embarked with Israel on the controversial invasion of the Suez Canal, in an attempt to prevent its nationalization by Egypt's President Nasser. The consequent closure of the canal cut a key transport lane – along which oil from the Middle East had been reaching Britain. Suddenly recognizing the vulnerability of its oil supplies, and still worried about an anticipated shortfall in the supply of domestic coal, the British government announced in March 1957 a revised and expanded nuclear programme. The new programme was aimed at having 6000 megawatts of nuclear power in service by 1965. Despite the increase in unit size of station – twin 275-megawatt reactors at Berkeley, twin 500-megawatt reactors at Hinkley Point – the expanded programme implied perhaps twelve stations.

The consortia were heartened by this increase in their expectations; but the gratification was short-lived. In the following two years the price of oil did not rise but fell, and the supply of coal from British mines steadily mounted. The latest generation of coal- and oil-fired generating stations, with ever-larger boilers and generating sets, were achieving better and better efficiency; their capital cost per kilowatt of output was showing a marked decrease, making the Magnox nuclear stations – indeed the very concept of gas-cooled reactors fuelled by natural uranium – look less and less attractive.

On 1 January 1958 the Central Electricity Generating Board, created by the Electricity Act of 1957, took over responsibility for generation of electricity 'in bulk' – a curious and inapt phrase – for England and Wales. The first chairman of the CEGB was Sir Christopher (later Lord) Hinton, the brilliant engineer who had overseen the creation of Britain's nuclear-weapons facilities. By putting him at the helm of the new CEGB, the government and its advisers might have been hoping to smooth the way for a rapid commitment to nuclear electricity generation. If so they reckoned without Hinton's own stubborn commitment to sound engineering – including engineering economics. Although Hinton had been a Member – that is, a board member – of the AEA, he had no illusions about the true economic status of nuclear power at the end of the 1950s. He was prepared to build and operate nuclear stations on the CEGB system, in order to acquire experience and continue the development process; but he did not pretend, or allow others to pretend, that these stations were a plausible economic alternative to contemporary fossil-fired stations. As a result, Hinton was soon at odds with his erstwhile colleagues in the AEA. One immediate controversy centred on the March 1957 revision of the first nuclear programme.

In 1959, after a worrying wait, yet another consortium had come to the head of the queue. The fifth in line, Atomic Power Constructions, led by International Combustion and Fairey Engineering, got the order for the Trawsfynydd station in north Wales. However, by this time the nuclear community could no longer defend its earlier insistence on an expanded nuclear programme as insurance against interruption of oil supplies. In early 1960, as the order picture grew progressively bleaker, the first two consortia, the Nuclear Energy Company and the Nuclear Power Plant Company, joined together as The Nuclear Power Group (TNPG). In June 1960 the Minister of Power, Richard Wood, told the House of Commons that the nuclear programme was being 'extended' – a euphemism for 'delayed'. It would now aim to have 5000 megawatts in operation by 1968. Wood insisted that this was merely 'a deferment of the acceleration which was planned in 1957', a form of words that fooled no one. The consortia found this to-ing and fro-ing about the scope of the programme scarcely encouraging. Their initial enthusiasm rapidly subsided. By late 1960 the Atomic Energy Group had merged with Atomic Power Constructions, in an uneasy partnership taking the name of the United Power Company. It was not, however, to remain long united.

When the Central Electricity Authority ordered the Berkeley and Bradwell stations in December 1956, they were expected to be completed and in service by 1961. They were, however, twelve to fifteen months late, a schedule overrun of some 20 per cent. The Hunterston and Hinkley Point stations in turn were about two years late, establishing a trend that would later become dramatic. As it happened, the capacity of the stations was not needed; but the delays meant that the CEGB and the South of Scotland Electricity Board (for Hunterston) had to run stations with much higher fuel costs, while continuing to pay interest on the capital tied up in the unfinished Magnox stations.

Work started in 1960 on the CEGB's fifth Magnox station, at Dungeness in Kent, and in 1961 on the sixth at Sizewell in Suffolk. But the Magnox design was proving to be so costly, in materials especially, that it looked less and less plausible as a basis for subsequent nuclear stations. A Magnox reactor produced comparatively little heat per unit volume of the reactor. Accordingly, achieving larger output required a major increase in the physical size of the reactor. Each successive Magnox station incorporated a pair of reactors larger than any of those in earlier stations. The core of a Magnox reactor was enclosed in a welded steel pressure vessel to confine the cooling gas used to collect the heat from the chain reaction. With the progressive size increase, welding such an enormous vessel out of steel plate became prohibitively difficult, especially because the welding had to be of the highest standard; rupture of a reactor vessel would lead to an extremely nasty accident.

Accordingly, the eighth commercial Magnox station – the CEGB's seventh, at Oldbury-on-Severn – incorporated two reactors whose pressure vessels were made not of steel but of prestressed concrete.

Concrete was a much easier material to work with on the necessary scale. It did, however, have certain drawbacks, as the builders of the Dungeness B AGRs later found to their cost. The ninth Magnox station, to be located at Wylfa on the island of Anglesey in north Wales, pushed the design to its limit – or possibly slightly beyond. In 1965 the Wylfa station was due in service by 1968; in the event, it did not even start up until 1971, and was still being 'commissioned' until 1975. Each of the two reactors was housed inside a vast spherical concrete pressure-cavern some 30 metres in diameter: vast reactors, and vastly expensive. Even at the time of the Wylfa order, in 1963, nuclear planners knew they could go no farther along the Magnox route. The question had long since arisen: what next? What design of reactor could take over from the Magnox concept, and become the basis for the next generation of nuclear stations?

Even in the 1940s British nuclear designers had come up with a plethora of possible design concepts for reactors. While working with the various consortia on continual revisions and up-dates of the Magnox design, the AEA was already pursuing development of not one but four other types. They included heavy-water reactors, high-temperature reactors, and fast breeder reactors, of which more later. The design of immediate interest, however, was a first cousin to the Magnox design. Like the Magnox design it used carbon dioxide as coolant and solid machined graphite blocks as so-called 'moderator' to facilitate the chain reaction. Its fuel, however, was not uranium metal but the much more durable ceramic uranium oxide, encased in cladding of stainless steel. Since the stainless steel absorbed rather a lot of neutrons the uranium had to be 'enriched', to increase the percentage of fissile uranium 235 in it. The design was called the advanced gas-cooled reactor, or AGR.

In 1957 the Authority got the government go-ahead to build a pilot plant of this design. It was to have a heat output of 100 megawatts, producing in turn some 27 megawatts of electricity, and was to be located on the northern edge of the AEA complex at Windscale, on the coast of what was then Cumberland. The Windscale AGR was designed, constructed and commissioned between August 1957 and December 1962. It reached full power of 100 megawatts of heat in January 1963, and was supplying electricity to the national grid in February 1963. In due course the AEA found that its output could even be raised, to 33 megawatts. The Windscale AGR was to be sure a pilot plant, and not intended to be an economic power station; but it performed impressively. With brutal hindsight it might have been better had the Windscale AGR not been so technically successful. Its success paved the way for a truly spectacular subsequent fiasco.

As noted above, the Magnox programme had been decreed by the government, at the instigation of the AEA. The CEGB, which had to carry it through, came formally into being only after the programme was well underway. By 1960, nevertheless, the CEGB, under its dynamic chairman Sir Christopher Hinton, was gradually developing a mind, and a nuclear policy, of its own. Hinton was instrumental in bringing about the cutback of the programme announced in the White Paper of June 1960. At the time, and until the spring of 1961, Hinton indicated that the CEGB was looking with favour on the AEA's AGR: that indeed an order might soon be forthcoming, as a follow-on from the Magnox programme. Then, in December 1961, Hinton published in the *Three Banks Review* a paper that badly shook the AEA.

In the paper Hinton reflected that different countries had followed different pathways into civil nuclear technology, and that it was not yet clear which would achieve success. He noted that to achieve such technical successes it was not necessary for a country to invent everything involved: some parts might with advantage be acquired under licence from elsewhere. The last sentence of the paper made Hinton's message unambiguous: 'Ultimately everyone connected with the development, design and construction of nuclear power plants must decide his research and development programme on the basis of what his customers find most economical and what he can develop and sell to give him a profit and them power at the lowest possible cost.' The AEA construed this to mean that the CEGB was not entirely in sympathy with the AEA's reactor-development programme, or wholly sanguine about the AGR. At the time, serious questions remained about the behaviour of the AGR's graphite moderator under the severe conditions in an operating reactor. Hinton knew about the graphite questions, and noted in the *Three Banks Review* that using graphite presented 'grave difficulties' – that 'it might be necessary to abandon the use of graphite as moderator and use heavy water instead'. The AEA's heavy-water design was still in its infancy;

but the Canadian CANDU heavy-water reactor was farther advanced, and by 1962 the CEGB was evincing what the AEA regarded as an entirely unhealthy interest in the CANDU.

A further complication from across the Atlantic was also gradually entering the British nuclear picture. In the United States there had been affronted outrage that the British Calder Hall station had been the first 'nuclear power plant' to start up and supply electricity to the public system, beating the Americans at what they still considered their own nuclear game. In the mid-1950s the US Atomic Energy Commission had launched a Cooperative Power Reactor Development programme; but the multitude of different design concepts constructed demonstrated only that most of the designs had little to offer. Only two types of reactor, both with military antecedents as marine propulsion units, lasted the course. Both used ordinary water as the coolant. One allowed the cooling water to boil inside the core of the reactor; the other kept the water under a pressure of up to 150 atmospheres to keep it from boiling. American enthusiasm for acronyms at once shortened the boiling-water reactor to BWR, and the pressurized-water reactor to PWR. The BWR was the brainchild of the mammoth US corporation General Electric – no relation, it should be noted, to Britain's General Electric Company. The PWR was developed initially by Westinghouse, and subsequently also by Combustion Engineering and Babcock & Wilcox.

In 1962 the CEGB, while watching American activities with interest, was dubious about both water-cooled designs. In due course events were to change that posture; but only after a great deal of policy had flowed under the bridge. In the early 1960s the transatlantic challenge to the AEA came from Canada. The CANDU used natural, not enriched, uranium. In the eyes of the CEGB this was a distinct advantage, given the limits on enrichment capability in Britain. The AEA's heavy-water design called for enriched fuel, another drawback. Furthermore, a CANDU prototype would have been operating for a year before the AEA's Windscale AGR would start up.

In the light of the mounting tension between the AEA and the CEGB, the government in the summer of 1962 appointed a committee to look into the question of choice of reactor for future nuclear stations. The committee was chaired by Sir Richard Powell, permanent secretary at the Board of Trade. Because the Powell committee was, in an otherwise unspecified way, declared to be under the aegis of the Cabinet, its deliberations were held in secret, and its eventual report remained unpublished – much to the discontent of concerned MPs and other interested parties. These secret hearings and discussions were to become endemic in subsequent re-runs of the reactor-choice controversy. The secrecy did not, to be sure, enhance the quality of consequent government policy, but doubtless it spared a few profound blushes.

The question of choice of reactor was only one of many by this time arising. From March 1962 to February 1963 the House of Commons Select Committee on Nationalised Industries held hearings into the activities of the electricity supply industry. One particular focus of their inquiries was the state of play regarding nuclear power – although, ironically, the committee's terms of reference did not in fact cover the Atomic Energy Authority, which was in this respect as in so many others a law unto itself. Nevertheless, the committee quizzed civil servants and representatives from the Atomic Energy Authority as well as from the electricity boards and manufacturing industry. In respect to nuclear policy the answers they received, and their consequent comments, make fascinating reading two decades later. According to the committee's report, published 28 May 1963, the chairman of the Atomic Energy Authority, Sir Roger Makins,

> thought that the size of the present programme was 'about right' until nuclear power is established as competitive. A larger programme in his opinion would have put a considerable strain on available building resources . . . At the same time, he believed that a smaller programme, in which reactors were built intermittently or in smaller sizes, could not achieve the objective of producing electricity at a cost comparable with conventional stations at an early date.

That frank admission – that the Magnox reactors were neither competitive nor indeed economic – was put more strongly by Hinton. As the committee reported,

> The Chairman of the Generating Board agreed that one order for a nuclear power station should be placed about every year to keep the three consortia satisfactorily occupied . . . But he considered 'in the light of hindsight' . . . that, if the past history of development had been different and he had been planning a programme with the sole object of meeting the industry's need for

nuclear fuel in the 1970s, he would have aimed to build a new reactor every two or three years interposing perhaps a small model of a new type when advancing technology justified it . . . Bearing in mind the history and structure of the industry, the Generating Board do not favour any further reduction in the size of the present programme . . . But the cost of that history to the Board is said to have been 'pretty considerable'.

In his evidence to the committee on 9 May 1962 Hinton gave some specifics. 'The costs which are being achieved on Berkeley and Bradwell are well above the costs which were estimated when those stations were put in hand.' The original tender price for each station had been 'just under £150 per kilowatt' of output; by early 1962, just before the start-up of the stations, the cost had reached £167. Compared to cost escalation on later stations, a mere 10 per cent in five years was, to be sure, trifling; but it was on top of an initial estimate already totally uncompetitive with other types of available generating plant. Sir Dennis Proctor, permanent secretary at the Ministry of Power, conceded this point before the committee. As they reported, he agreed

that until nuclear and conventional power become competitive, the industry will be bearing the additional cost of generation because of 'national policy as laid down by the Government', although he claimed that the industry accept the additional cost now in order to gain the long-term advantage. He readily accepted the Board's figure of £20 million a year [additional cost of generating electricity by nuclear rather than conventional power stations] although it had never been specifically discussed with them. In his view the argument as to whether the taxpayer or the electricity consumer should bear the extra cost of the nuclear power programme should proceed from the basis that in ten or fifteen years' time nuclear power stations will be needed, and that just as present customers have benefited from technological advances in the past, so they should bear the cost of present advances. The witness did, however, agree that it was hard for the industry to be saddled with the extra cost [£360 million extra capital cost for seven nuclear stations between 1962 and 1968, compared to conventional stations] of a programme which is now generally admitted to be too big and which it is doubtful they would have supported if they had been 'perfectly free agents'.

One comment from Hinton encapsulated succinctly the state of affairs by this time prevailing between the CEGB and the AEA. One committee member, the splendidly-named Sir Henry d'Avigdor-Goldsmid, put a proposition to Hinton: 'But now that you [the

CEGB] are the main client of the AEA it seems fair to suppose that their activity is guided by your requirements?' Hinton replied tersely: 'I think that their activities are guided by what they think our requirements ought to be.' It was not the reply of a satisfied client.

The contracts for the first five Magnox stations had been awarded to five different consortia; after the five had shrunk to three the four remaining orders were split between two of them. As may be evident, the notion of competitive tendering, with contracts going to the lowest bidder, had never got off the ground. The contracts were awarded more or less according to the old civil service criterion of 'Buggins' turn', each consortium in rota getting the call. Even though the electricity boards were the customers, the choice of main contractor was never left to the boards alone. Whitehall and the AEA each had a prominent say in the matter, and the final outcome of each award was the result not so much of competitive tendering as of competitive lobbying and jockeying for position between the different groups.

By 1962 the collective construction capacity of the three surviving consortia had been scaled down substantially; but it was still much larger than the anticipated ordering programme could well support. Of the three survivors one in particular, the United Power Company, was in a precarious condition. It had expected to receive the contract to build the Wylfa Magnox station; but the contract had instead gone to the English Electric–Babcock & Wilcox group in circumstances that eventually led to accusations of bad faith, and a fierce debate in the House of Lords. As the Magnox programme ran out its course, the three consortia had an understandably nail-biting interest in the choice of reactor for the prospective second nuclear programme, not to mention the choice of who would build it.

The AEA was pressing ever more strongly for the advanced gas-cooled reactor, and all three consortia were using AEA data to develop working designs of full-scale AGR. But two of the consortia, mindful of the CEGB's reservations about the AGR, were also pursuing parallel development of water-reactor designs under licence from the US. Until late 1963 the US electricity suppliers, both publicly and privately owned, remained sceptical about the novel technology of nuclear power. They were prepared to build and operate nuclear plants only if someone else – the US government, that is, US taxpayers – footed the bill. In December 1963, however,

US General Electric announced the breakthrough that the nuclear world had been waiting for. Jersey Central Power and Light had ordered a BWR nuclear plant for its site at Oyster Creek, with no government subsidy whatever. The taxpayers would not have to contribute a cent.

The Oyster Creek contract was at once acclaimed around the world – not least in Britain – as a sign that nuclear power, on the US water-cooled model, was now cheaper than coal- or oil-fired electricity generation. Within a few months the nuclear orders were flooding on to the desks of the US reactor vendors. Only later did it become evident that Oyster Creek, and all the subsequent so-called 'turnkey' plants ordered at a fixed price, to be handed over ready to operate, were being sold as 'loss leaders' by the manufacturers. By 1976, according to the *Wall Street Journal*, they were to cost their suppliers some $2000 million in losses. Nevertheless, in Britain in 1964 the apparent success of the US water-cooled designs was seized upon as a sign that Britain too should adopt a water-cooled reactor as the basis for its second nuclear programme. The in-fighting between the three consortia and their supporting factions grew ever more intense, with the AEA vigorously pulling strings behind the scenes.

Long after the dust of the Select Committee report had settled, and after the Powell committee had apparently failed to resolve the question of reactor choice, the government at last produced its eagerly-awaited statement on future nuclear power policy. It consisted of a three-page White Paper, published in April 1964; and it said, as might be expected from its length, very little about the key issue occupying British nuclear minds. Instead of announcing which reactor would get the glass slipper, the White Paper announced only that the CEGB was to be instructed to call for tenders, inviting the consortia to offer designs based on either the AGR or US water-cooled reactors. The CEGB would then carry out a meticulous economic assessment of the competing designs, and make its choice according to the outcome of this assessment, on a purely commercial basis. Such, at least, was the story.

All three consortia duly submitted tenders. All three offered AGR designs on the basis of what was said to be even-handed information provided by the Atomic Energy Authority to all three. The Nuclear

Power Group, drawing upon its link-up with US General Electric, also offered a BWR design; the English Electric–Babcock & Wilcox group, drawing on its link with Westinghouse, offered a PWR design. Various constituent companies also tendered for various bits of the proposed station. However, even as the United Power Company was getting its AGR act together, it came untied. In early 1965 GEC pulled out, leaving Atomic Power Constructions and its constituent companies as the rump of UPC.

In December 1964 Lord Hinton retired from the CEGB. His successor as chairman was F.H.S. (soon to be Sir Stanley) Brown, who thereupon took over the job of choosing the lucky winner in the reactor stakes. The deadline for tenders was 1 February 1965. The CEGB thereafter set about its meticulous commercial assessment. The outcome, however, dumbfounded most informed onlookers. When at last, on 25 May 1965, Fred Lee, as Minister of Power, made what was to be the most famous – or notorious – speech of his career, the lucky winners were revealed to be not only the AGR but also Atomic Power Constructions. APC, still shaky from the after-effect of losing not only the Wylfa contract but also its strongest industrial partner in GEC, had apparently hit pay-dirt nevertheless. Ere long it was to become clear that hitting pay-dirt can mean falling flat on your face.

The CEGB was so proud of the economic assessment it had carried out that it published a short treatise entitled 'An Appraisal of the Technical and Economic Aspects of Dungeness B Nuclear Power Station' describing the method and the results, and circulated copies to anyone interested. For purposes of explanation the appraisal compared the cost of electricity from the proposed Dungeness B AGR offered by Atomic Power Constructions to that from the 'next most attractive offer', the US General Electric BWR offered by TNPG. These costs were in turn compared to those for the contemporary Cottam coal-fired station and the Wylfa Magnox station. The appraisal described the assumptions used, and arrived at generating costs of 0.457 old pence per kilowatt hour (d/kWh) for Atomic Power Constructions' AGR, and 0.489d/kWh for General Electric's BWR. Note the three decimal places. Nuclear hubris at its most extreme has rarely achieved the transcendence of the CEGB's Dungeness B appraisal.

Without going any further into an analysis of the assumptions that proved nonsensical, the difference of about 7 per cent in generating cost should be set against the so-called 'availability adjustment' charged against the BWR. Since the BWR had to be shut down for refuelling, the appraisal added about 6.5 per cent to the total generating cost, on the basis that the AGR would be refuelled on load, without shutting down. In the event no commercial AGR has ever successfully refuelled an entire core on load at power; even partial on-load refuelling has been achieved only in the 1980s. Correcting this single incorrect assumption alone would have left the appraisal essentially balanced on a knife-edge between the AGR and the BWR. It need scarcely be added that virtually every other quantitative assumption in the appraisal was similarly ill-founded – not least that which anticipated having Dungeness B 'on load in 1970'. In retrospect it is scarcely surprising that the CEGB in the 1970s and 1980s was so reluctant to spell out the thinking behind its many subsequent prognostications. The Dungeness B appraisal of 1965 is worthy of a place in the annals next to the euphoric pronouncements of the White Star Line just before the maiden voyage of the *Titanic*.

One far from incidental side-effect of the much-ballyhooed 'economic assessment' was to deflect attention, at least temporarily, away from what appeared to be some vigorous political string-pulling behind the scenes. The original specification for the AGR as laid down by the CEGB called for fuel-elements of a particular design. The other two consortia complied with this stipulation; but Atomic Power Constructions based their design on a more complex element, still at an early stage of development by the Atomic Energy Authority. The AEA bridled at the suggestion that they had given APC data not available to the other consortia; but discontented murmuring continued. The impression persisted that the AEA, alarmed at the prospect of their own AGR being underbid by the foreign interlopers, had colluded with APC to come up with design parameters that were at the limit of engineering knowledge, if not indeed beyond it.

This allegation was indignantly denied by the AEA at the time, and even more vehemently later, as the débâcle at Dungeness grew grimmer. Certainly the basic difficulty arose in the original CEGB specification, which called for reactors capable of generating 600 megawatts each – some twenty times the size of the only operating AGR, the 30-megawatt pilot plant at Windscale. Indeed, when the formal order for Dungeness B was placed, in August 1965, it called for a twin-reactor station with a design output of 1320 megawatts. A scale-up so dramatic was asking for trouble. It duly arrived.

Before it did, however, the CEGB had ordered another AGR station and the South of Scotland board had ordered its first, both from The Nuclear Power Group. They were to have design outputs of 1320 megawatts and 1330 megawatts respectively; each would be built on a site next to an existing Magnox station. The new stations, known as Hinkley Point B and Hunterston B, were both ordered in 1967, and construction was underway before the end of the year. It was to last a good deal longer than either board expected.

By this time another significant participant had joined the fray. The Labour government, in the first flush of Harold Wilson's 'white heat', brought into being a new Parliamentary body: the House of Commons Select Committee on Science and Technology. It was an all-party committee of backbench MPs with an interest in the social and political dimensions of their subject. British nuclear power was an ideal topic for their inaugural investigation. It offered just the sort of political battleground on which backbenchers could have a field day: complex specialized information, hotly competing interest groups, large sums of money changing hands, and a clear-cut requirement for oversight by the tribunes of the people. The committee was created in December 1966; its members were nominated in January 1967; they set to work in February; and their hefty report was published in October. In the course of its preparation they took evidence from an array of eminent witnesses, including AEA chairman Sir William Penney, CEGB chairman Sir Stanley Brown, the heads of the three reactor-building consortia, the Minister of Technology, Anthony Wedgwood Benn as he was then known, the Minister of Power, Richard Marsh, and – something of a spectre at the feast – Lord Robens, chairman of the National Coal Board.

Robens appeared before the Select Committee in defence of his own beleaguered industry. The government was drafting its second White Paper on Fuel Policy in only two years; but even in those two years the picture for coal had grown distinctly blacker. The CEGB's 'economic assessment' of Dungeness B had apparently convinced the

Generating Board that its earlier reservations about nuclear power could be set aside. The departure of Lord Hinton probably helped. As a result, the CEGB was by 1967 proposing to order its third AGR station – for a site on the very edge of the Durham coalfield, at Seaton Carew. Board officials insisted that AGR electricity would be cheaper than coal, even virtually at the mouth of the mine. Were this so, the Coal Board and its industry would be faced with inevitable extinction. The electricity boards were already by far the largest users of coal, as manufacturing industry converted to oil. If even the electricity boards were no longer prepared to buy coal by the many millions of tons, the one-time black diamonds had no future.

This gloomy conclusion was underlined by the government White Paper on Fuel Policy, published in November 1967. It accepted without question that the future of fuel supply lay with oil, gas and nuclear power; the outlook for the coal industry was to be gradual and orderly euthanasia. Lord Robens was, however, nothing if not a fighter. In answer to the CEGB's proposal to site an AGR station at Seaton Carew, Robens offered the CEGB coal at a bargain rate. The CEGB was unmoved. After a brief ministerial delay the AGR plan went ahead, with the site now called Hartlepool.

The Hartlepool site, as well as being on the doorstep of coal country, was also in the middle of the industrial complex of Teesside – closer to a major urban area than any previous nuclear station site. In the eyes of the Nuclear Installations Inspectorate the move to prestressed concrete pressure vessels, with their enhanced safety factor, made such near-urban siting acceptable. Ere long, however, the Nuclear Installations Inspectorate was to have some expensive second thoughts.

One of the little-known casualties of the Seaton Carew–Hartlepool decision by the CEGB was Britain's world leadership in the advanced coal technology called 'fluidized-bed combustion' – FBC for short. While the Coal Board and the CEGB were at loggerheads, the Coal Board Member for Science, Leslie Grainger, was overseeing the planning of a proposal for a 20-megawatt prototype FBC power station. It would be sited at Grimethorpe colliery in Yorkshire, using coal from the pit-head, and demonstrating the possibility of burning it cleanly and efficiently with minimal pre-treatment. Grainger was keen to build the station with Coal Board research funds; but Robens insisted that the CEGB must be persuaded to put up money, to ensure that it would thereafter remain committed to further development of this advanced coal technology. The CEGB, however, was loftily uninterested in advanced coal technology; its own work in the field had been wound down by the mid-1960s. The Grimethorpe FBC plan had to go back on the Coal Board shelf. Not until 1984, after many vicissitudes, did the CEGB at last join the Coal Board at Grimethorpe – by which time FBC was becoming internationally recognized as the most desirable way to burn coal and control air pollution. Unfortunately, the lead in FBC technology held by Britain in the late 1960s had long since passed to Scandinavia and the US. The first reactor at the Hartlepool nuclear station did not even start up until 1983, sixteen years later; not until 1985 were both Hartlepool AGRs at last supplying electricity to the grid.

The first report from the Select Committee on Science and Technology appeared on 25 October 1967. Its first recommendation was that the consortium system be 'phased out', and it spoke up in favour of a single manufacturer of what it called, with perfect correctness but an oddly archaic ring, 'nuclear boilers'. A like recommendation had come from the Select Committee on Nationalised Industries in its report in 1963; but the difficulties remained. Not only were the remaining three consortia on far from cordial terms; one of them, Atomic Power Constructions, was slowly sinking into deep trouble on the south Kent coast. Neither of the other two consortia much liked the idea of being in bed with Dungeness B.

The problems at Dungeness B revealed themselves gradually. As late as 1968 the managing director of APC, Gordon Brown, was calling the AGR 'the only system with true development potential'. Unfortunately, it was also a system with quite staggering potential for developing trouble. Part of the difficulty stemmed from the hectic winter of 1963–4 in which APC put forward its winning bid for the contract. Pressure of time meant that the design was still at best preliminary when work on site got underway; and the aim of having the plant on line by 1970 left the designers making it up as they went along.

They were trying to get three times as much power as the adjoining Dungeness A Magnox station out of a plant one-third the volume; and the tolerances they adopted left little room for error – too little room, as it proved, quite literally. The prestressed concrete pressure vessel had to be lined with steel. The steel liner was distorted during construction, and further distorted during installation – so much so that the boilers, intended to be inserted in chambers in the pressure vessel, did not fit. In due course the entire upper half of the liner walls had to be dismantled and rebuilt. In turn, the original design of boiler had been found by 1968 to be impracticable; the casings, hangers and tube supports all required redesign. The redesign failed to resolve one key problem that subsequently dogged not only Dungeness B but also its sister plants at Hinkley Point B and Hunterston B. Under operating conditions, the carbon dioxide coolant in these reactors was scarcely recognizable as a gas. At the operating temperature and pressure it was more like a liquid; and as it rushed through the fuel channels and associated pipework in the reactor it pounded and hammered the fittings, setting up vibrations that threatened to tear the system apart. One redesign after another failed to overcome this gas-vibration problem, not only at Dungeness B but also at the next two plants.

These technical problems were compounded by problems of finance, management and staffing, occasioned not least by the increasing likelihood that Atomic Power Constructions would never receive another order for anything from anybody. While APC sank inexorably beneath the waves its erstwhile partner, GEC, completed a merger with English Electric; and in late 1968 the reshuffled consortium re-emerged as British Nuclear Design and Construction (BNDC). After many fraught months BNDC agreed to take over project management of Dungeness B. APC's member-companies agreed to pay £10 million compensation to the CEGB; APC was formally reduced to a rump of a company, to remain in existence only until Dungeness B was completed.

Yet another problem with the gas-cooled designs surfaced in 1968, initially with the original Magnox reactors. While the design had been under development, the nuclear engineers had been concerned about the possibility of chemical reactions between the carbon dioxide and the reactor internals, including the graphite moderator and the fuel cladding. In almost every part of the reactor core they specified materials that had been found experimentally to be safe from corrosion by hot carbon dioxide. Unhappily they overlooked certain bolts; and after the early Magnox plants had been in operation for a few years, these bolts, stressed by accumulating corrosion, showed signs of breaking. By this stage the only available remedy was to lower the operating temperature in the reactor cores, to slow down the corrosion. The reduction in core temperature led to a corresponding reduction in heat output from the plants. This 'derating' affected all but one of the Magnox plants, becoming the more severe the larger the plant. In due course the Oldbury plant, whose design output had been 600 megawatts, was to be derated to only 415. The Wylfa plant, whose troubles in the late 1960s were only beginning, was eventually derated from 1190 megawatts to a mere 840 – two reactors for the price of three.

It should be added that the nuclear plants were by no means the only headache afflicting the electricity boards in the late 1960s. The headlong scale-up in size of plant was common not only to Magnox and AGRs but also to coal- and oil-fired stations, and to their turbo-alternator sets. No fewer than forty-seven sets with outputs of 500 and 660 megawatts had been ordered since the early 1960s, before even a single set of this size had run; and virtually every set gave problems. Site work suffered from abysmal labour relations, with sudden strikes, poor quality control and embarrassing productivity – so much so that the Labour government in July 1968 appointed a Committee of Inquiry under Sir Alan Wilson, FRS, to investigate 'Delays in commissioning CEGB power stations'. The committee duly reported in March 1969; but the situation at CEGB power station sites remained fraught into the 1980s. To add to the CEGB's travails, the Nuclear Installations Inspectorate declared in 1970 that it was belatedly unhappy with part of the design of the boilers for the Hartlepool AGR station. The consequent modifications set the CEGB back some £25 million, and the project back yet further.

The AEA, it must be said, was not particularly moved by the travails of the AGRs. Instead it pressed busily on with yet more reactor designs, with no apparent strenuous thought about what roles they might conceivably play in the real world of generating electricity for paying customers who would not want to go on indefinitely paying over the odds. The AEA was by this time operating the 100-megawatt prototype of its own steam-generating heavy-water reactor (SGHWR), at the AEA's site at Winfrith in Dorset; the Winfrith AGR had started up in 1967. The AEA's pet project, the fast breeder reactor, was well into its second phase with construction of the Prototype Fast Reactor in progress next door to the Dounreay Fast Reactor along the coast from John o'Groats. The AEA was also becoming the prime mover behind the 20-megawatt (thermal) Dragon experimental high-temperature reactor (HTR) at Winfrith. Originally established as a multinational nuclear research and development project by the Organization for Economic Cooperation and Development (OECD), the Dragon reactor was to undergo a painful and ultimately terminal technological identity crisis. At the beginning of the 1970s, however, no one had anything but good to say about it – perhaps an ominous sign in the circumstances.

By 1970 plans were well on the way for the AEA itself to undergo fission. In June 1970 Harold Wilson called a general election, and lost it. With it went the pending legislation to reorganize the AEA. But the incoming Conservative government of Edward Heath reintroduced the legislation immediately, and the break-up duly took place in 1971. The Atomic Energy Authority Act received the Royal Assent on 16 March 1971, splitting the monolithic Authority into three constituent parts. On 1 April a new company called the Radiochemical Centre Ltd took over production and marketing of medical and industrial radioisotopes; and another new quasi-'commercial' company, called British Nuclear Fuels Ltd (BNFL), took over all the AEA's fuel-service activities, including the manufacturing plant at Springfields, the enrichment plant at Capenhurst, the spent-fuel complex at Windscale, and the dual-purpose Calder Hall and Chapelcross plutonium-plus-electricity reactors. The ostensibly 'commercial' nature of BNFL was slightly shadowed by the fact that 100 per cent of the shares in BNFL were held by the AEA, and that AEA chairman Sir John Hill was also installed as chairman of BNFL.

According to BNFL's first annual report,

> The principal activities of the Company are:
> a) Nuclear Fuel Services
> The conversion and enrichment of uranium; the manufacture of uranium and plutonium based fuels and the provision of related fuel cycle services for nuclear power stations; and the reprocessing of nuclear fuel after use.
> b) Electricity
> The operation of two nuclear stations for electricity generation.

In addition the Company manufactures specialised components from depleted uranium and from graphite, undertakes irradiations in its nuclear reactors, and prepares radioactive substances. It also engages in research, development and the design and construction of plant and equipment associated with the principal activities.

It was a curiously coy pronouncement in a crucial respect. No mention was made anywhere in the report, or in its successors, of the activity for which the facilities hived off to BNFL had been originally constructed. Only one tell-tale glimmer could be glimpsed. Buried in the fine print of the 'Notes on the Accounts' was this intriguing sentence: 'Assets originally provided for Defence [capital in original] purposes and which the Company may in certain circumstances be required to use for such purposes had no value attributed to them on their transfer to the Company.'

It suggested an image of the hapless directors of BNFL being 'required' by an implacable superior authority to use its 'assets' for dark and mysterious purposes. But of course all it meant was that this 'commercial' company continued to manufacture plutonium for nuclear weapons in the reactors at Calder Hall and Chapelcross, and to recover and process this weapons-plutonium in the facilities at Windscale, as the AEA had done before BNFL was even a gleam in official eyes. BNFL remained from its inception onwards essentially mute about this key activity, so much so that even in 1985 few people are aware of it: a remarkable state of affairs, especially in the light of government plans to sell shares in BNFL to private investors. Will the prospectus state that the company is the purveyor of weapons-plutonium to Her Majesty's Government; or will it maintain the discreet silence that has prevailed since 1971?

The break-up of the AEA did little to ease the steadily mounting confusion in British nuclear power policy. Accordingly, while the Heath government was dismantling the AEA it was also trying to get to grips with the ever more convoluted problem of reactor choice. The CEGB in 1970 had ordered its fourth AGR station, to be sited at Heysham on the curve of Morecambe Bay in Lancashire. But all the earlier AGRs were giving rise for concern, as one technical hitch led to another, schedules slipped and costs escalated. The PWR supporters looked on sardonically. Sooner or later, they were certain, Britain too must go the water route.

Confronted by this stubborn dilemma the Heath government did what governments usually do about dilemmas: it set up another committee. The committee was chaired by Peter Vinter, a deputy secretary in the Department of Trade and Industry. The committee's brief was to weigh the corrosive question of reactor choice yet again, and report. It need hardly be said that the Vinter committee was to operate, as usual, in secret; its report would be for the insiders only. As usual, the secrecy served to conceal official embarrassment: because the Vinter committee apparently found its brief impossible. When at length, in 1972, its report was delivered, so far as could be determined it made no attempt to recommend choosing one reactor or another. Instead, it apparently returned to another similarly chronic and corrosive issue: the need to streamline Britain's reactor-building industry, and slim it down to match the economically plausible demand on it.

Unfortunately, the two remaining consortia, The Nuclear Power Group (TNPG) and British Nuclear Design and Construction (BNDC), had no great fondness for each other, and showed little enthusiasm for the shotgun marriage proposed. Nevertheless, on 8 August 1972, John Davies, the Minister for Trade and Industry, read the banns. There would be, he told the House of Commons, a new single company set up to build all of Britain's power reactors. On 22 March 1973 Davis's successor, Peter Walker, dotted the 'i's and crossed the 't's. The new company was to be called the National Nuclear Corporation, or NNC for short. NNC in turn would control a subsidiary called the Nuclear Power Company, which would do the actual design and construction work. Fifteen per cent of the shares of NNC would be held by the Atomic Energy Authority and 35 per cent by British Nuclear Associates, a group of seven companies left over from the old consortia. The other 50 per cent of NNC would go to Sir Arnold Weinstock's aggressive and predatory GEC; and GEC would take over effective managerial control of NNC. The government for its part would receive advice from a new Nuclear Power Advisory Board, drawn from the top echelons of the industry.

If the government thought its new plan would stamp out the internecine warfare in the nuclear corridors of power, it could not have been more mistaken.

2 Water pressure

'Until after 1980 we shall need to order at most four nuclear stations. On a more reasonable but still optimistic assumption about the growth of demand for electricity, we shall in that period need to order only one – and that one probably a fast reactor.' This – boiled down into two sentences – is what the new chairman of the Central Electricity Generating Board, Arthur Hawkins, told the Select Committee on Science and Technology on 2 August 1972. Sixteen months later the same man told the same committee that the CEGB now proposed to order as a matter of urgency some eighteen new nuclear stations, each twice the size of the largest hitherto built in Britain, by 1982. Even in the topsy-turvy world of British nuclear power policy, this extraordinary somersault stands out. It helped to provoke the first public awareness of the chaotic turmoil by now endemic behind the scenes.

Arthur Hawkins, deputy chairman of the CEGB, had taken over the chairmanship from Sir Stanley Brown on 1 July 1972. Hawkins first appeared before the committee on 2 August 1972, during its third investigation into British nuclear policy. The committee, it should be noted, was empowered to take up any suitable aspect of 'science and technology' with a Parliamentary dimension. Its preoccupation with nuclear policy dominated its concerns for the twelve years of its existence; and its effective successor, the Select Committee on Energy, set up in 1979, likewise got underway with yet another inquiry into nuclear policy. A dispassionate bystander is compelled to conclude that Members of Parliament were not entirely satisfied with the formulation and execution of nuclear power policy in Britain. The MPs had good reason for discontent. The appearances of Arthur Hawkins as a witness in August 1972 and then again in December 1973, during his hegemony at the CEGB, offered choice examples of the nuclear power brokers at their worst – arrogant, patronizing and egregiously inaccurate.

The Select Committee published its third nuclear report, entitled *Nuclear Power Policy*, in June 1973. It welcomed the government's 'decision to consolidate the industry into a single unit as at last implementing the Committee's recommendation of some five years ago'. It is possible to doubt whether the government saw itself as 'implementing the Committee's recommendation'. The committee, like the other brawling participants in the long-running wrangle, tended to see itself as the fount of nuclear knowledge, whereas successive governments saw it more as an incidental and somewhat irksome inconvenience – witness the refusal to let the committee see the Vinter report. Despite its satisfaction at the new single company, however, the committee was unhappy at the prospect that GEC might take a 50 per cent shareholding in it. 'Should GEC ever decide that they were no longer able to participate in the nuclear reactor industry, their withdrawal from the new company would have disastrous effects on its stability.' The committee's comment in due course proved all too prophetic.

The committee also saw little advantage in the government's proposed Nuclear Power Advisory Board. 'It is not unknown for advisory boards on scientific and technological matters to make reports which are totally ignored,' noted the committee with heavy irony. 'Should the Board not be given some guarantee of its key position in the scheme of things . . . in practice most decisions will, we suspect, be taken by the new company in collaboration with the CEGB and Ministry officials.' The committee's suspicions were if anything too circumspect. Even in a context where futility was rampant, the Nuclear Power Advisory Board was to prove to be in a class by itself.

One of the underlying reasons for its futility had been foreshadowed in the evidence given to the committee in early 1973 by the two old consortia. Witnesses for British Nuclear Design and Construction had made it clear to the committee that their longer-term preference was for the high-temperature reactor. The Nuclear Power Group, on the other hand, had much preferred the steam-generating heavy-water reactor – understandably, since it had built the 100-megawatt prototype at Winfrith. BNDC, with GEC in the driver's seat, had been bullish about the prospects for light-water reactors, not only in

Britain but also for export. TNPG had had certain reservations, as had the Chief Inspector of Nuclear Installations, Eric Williams.

Stoutly defending his independence in his first appearance before the committee, on 20 February 1973, Williams had played a tight hand. As the committee ruefully noted, 'We questioned Mr Williams on the safety of light water reactors very closely, but without much success. He took the view that since he would be called upon to give his opinion on any application for a licence to operate such a reactor, he could not be seen to be prejudging the issue by expressing any opinion on the reactor's merits.' Williams's next appearance before the committee, only ten months after the first, was to be more revealing, if even less satisfying.

'It is generally agreed,' said the committee, 'that the consortium principle has not been a success, and the impression we gained from witnesses was that they wanted instead a company with strong commercial management, run as an entity in itself without undue interference by the various organizations holding shares in it. There was no general agreement on how to achieve this,' the committee added drily, in a masterpiece of understatement.

There was, to be sure, no shortage of excellent reasons for considering erstwhile nuclear power policy less than a howling success – Dungeness B not least. But the official decision to abandon the pretence of 'competitive' tendering by more than one supplier was dictated above all by the scarcity of forthcoming orders. CEGB chairman Arthur Hawkins had made this brutally clear on 2 August 1972.

If we take 5 per cent per annum [as the rate of growth of electricity demand] we should require between now and ten years ahead an additional 31,500 megawatts of plant. Of that, 15,500 are already authorized and committed. Therefore we only need to order in that period of ten years, on the basis of a 5 per cent growth, an additional 16,000 megawatts of plant for commissioning in that period. That would mean starting ten or eleven new stations, of which we would probably suggest that four would be nuclear stations, one probably a fast reactor. But that is on the basis of a 5 per cent growth. If we take the 3½ per cent per annum growth – which, as I remind you, is a little higher than the average over the last three to four years – we should only require 19,500 megawatts of additional plant, of which 15,500 megawatts have already been committed. Therefore we would only require 4000 megawatts more plant. That would mean in this period for commissioning until after 1980 only three new station starts, of which one would probably be nuclear – and probably the fast reactor.

It would be difficult to sustain even a single nuclear-plant company on such a starvation diet. The logic of the decision to create a single company depended above all on this paucity of foreseeable business. This was the main reason why on 22 March 1973 Peter Walker, Secretary of State for Trade and Industry, had told the House of Commons that the government was to create a single reactor-building company. As indicated in the preceding chapter, it would be called the National Nuclear Corporation. The government, through the AEA, would take 15 per cent of the shares, and a group of seven survivors from the old consortia would take jointly another 35 per cent. GEC, as rumoured, would take 50 per cent of the shares, and take over also as the management company under a separate contract.

Almost as soon as GEC had achieved the ascendancy, something very strange happened in the nuclear smoke-filled rooms. The first public inkling that curious and ultimately breath-taking developments were afoot came with a front-page article in the *Guardian* on 15 October 1973. The paper's astute energy correspondent, Peter Rodgers, reported that the CEGB was intending to abandon British gas-cooled reactors, and ask the government for permission to build American-designed pressurized-water reactors instead.

The timing probably had a good deal to do with the row that swiftly blew up. The Arab–Israeli Yom Kippur war was already threatening international oil supplies, and the Organization of Petroleum Exporting Countries (OPEC) was at last exerting its economic leverage on its overseas customers. In Britain, North Sea oil was on the verge of coming ashore for the first time, and Britons were being told that they too would soon qualify for OPEC membership. The coal miners, however, were growing restive, and the possibility of the second damaging coal strike in two years was looming. Against this background of front-page news, 'energy' had suddenly become a popular catchword. Partly as a result of heightened public awareness of energy issues in general, the latest convulsion in British nuclear policy attracted public attention – the first time a civil nuclear issue had ever done so. Previous nuclear power controversies in Britain had, to be sure, been ferocious and bitter; but they had taken place essentially behind the scenes, between the immediately interested parties and their supporters. This time the controversy came into the open; and the public noticed.

Another contributing factor was the rise of 'the environment' as a matter for popular interest and concern. In Britain the traditional wildlife and countryside organizations had been joined by a new breed of 'environmental' organization with a more aggressive approach to environmental issues. One such organization was Friends of the Earth (FOE), which by 1973 had become well known as a result of a series of highly visible campaigns. FOE's sister organizations in the US, Sweden and France were already embroiled in nuclear power controversies in their own countries. The CEGB decision to adopt light-water reactors triggered a campaign by British FOE to stir public debate about the proposal, and provide the information to fuel such a debate. British newspapers, radio and television had already begun to describe civil nuclear confrontations arising in the US and elsewhere. When a juicy confrontation happened right under their noses they seized upon it with alacrity. Before the end of 1973, British nuclear power policy was on the front pages.

The Nuclear Power Advisory Board had been duly constituted in August 1973. It numbered ten 'wise men', including the chairmen of the AEA, NNC, electricity boards and Electricity Council, with the Secretary of State for Trade and Industry in the chair. Its deliberations got underway immediately – in secret, needless to say. On 5 and 21 November and 4 December it was reported to have received papers and heard presentations from the interested parties about the latest nuclear intentions; the reports reinforced the belief that the PWR had leapt into the lead. On 20 November the CEGB filed an application to site a nuclear power station at Orford Ness, on the Suffolk coast. The outline application listed the possible reactors to be used as AGRs, PWRs, SGHWRs or HTRs. If it had also included Magnox it would have been a Concise roll-call of the factions already arm-wrestling in the corridors. On 26 November another country was heard from: Lorne Gray, chairman of Atomic Energy of Canada Ltd, visited London to sing the praises of his company's CANDU heavy-water reactors. His claims were echoed by British CANDU enthusiasts, not least the Select Committee on Science and Technology.

The committee, whose report on nuclear power policy had appeared only four months earlier, was already returning to the fray. On 13 November a deputation from the Energy Resources Sub-Committee of the Select Committee called on Energy Secretary Peter Walker to express their disquiet about the rumoured switch to the PWR. Walker reportedly told them that there would be no decision until January 1974; but they drew little comfort from the Minister's response, and forthwith announced that they would carry out yet another investigation of the subject.

On 29 November the government responded brusquely to the Select Committee's June report, with a four-page White Paper that briskly dismissed virtually every committee recommendation. This did not improve the committee's temper. After the government's brush-off of its report, with its strictures about giving GEC 50 per cent of the National Nuclear Corporation, and about the safety problems of light-water reactors, the committee was in a pugnacious mood. Between 11 December 1973 and 30 January 1974, the committee's Energy Resources Sub-Committee held six hearings, and drafted, agreed and published a crackling report taking curt issue with much of what its official witnesses had said in evidence. The witnesses included Sir Arnold Weinstock of GEC; Francis Tombs of the South of Scotland Electricity Board; Arthur Hawkins of the CEGB; Eric Williams of the Nuclear Installations Inspectorate; Lord Aldington, recently appointed chairman of the National Nuclear Corporation; and Sir John Hill, chairman of the Atomic Energy Authority. They got a pretty rough ride, in some cases deservedly so.

Opening the first sitting, on 11 December 1973, chairman Arthur Palmer made his feelings unambiguous. The first witness to appear was Sir Arnold Weinstock of GEC, now the managing director of the newly-fledged National Nuclear Corporation.

Since you came in front of us before, Sir Arnold, quite a lot has happened in the nuclear reactor field. The Select Committee did produce its report. The Government has made several announcements and a new nuclear manufacturing company has been set up. There has been a White Paper issued by the Government in reply to the Select Committee, rejecting almost entirely every one of our suggestions, except one, but we are used to that kind of rebuff.

In this and subsequent sittings Palmer and his colleagues then set about trying to ascertain why key witnesses were now contradicting what they had told the committee only the previous year, or indeed only the previous spring.

In March 1973, nine months earlier, Weinstock had told the Committee that although the British choice of gas-cooled reactor had not worked out well, he would not suggest that in the future Britain should move to water-cooled reactors. The large stake already committed to advanced gas-cooled reactors and the safety questions connected with the light-water reactor were, he felt, substantial reasons to stay with the British gas-cooled reactor. Confronted with this view nine months later, Weinstock alluded vaguely to the 'urgency' of the 'general energy situation'.

> After some talks with the CEGB and contemplating their programme and what they needed, not what they would like to have but what they absolutely have to have in the way of nuclear energy . . . I was obliged to modify my opinion and to arrive at the conclusion that the AGR does not offer anything like the necessary security of supply.

Sir Arnold went on to sing the praises of the light-water reactor, as the design most widely adopted outside Britain. 'If the larger part of the world is wrong in using light water reactors, very fundamental problems arise about the economic and industrial problems of the Western world.' Subsequent events were to prove that Sir Arnold spoke truer than he knew.

Two days later Francis Tombs of the South of Scotland Electricity Board (SSEB) took the stand. Tombs shared Weinstock's doubts about AGRs. With the Hunterston B plant already twenty-two months late, Tombs declared, 'Certainly we would be reluctant to order further AGRs in the present climate.' However, Tombs continued, 'Many views have been expressed about the American [light-water] reactors, and it is right to say that we are a little less sanguine about their choice than other views which have been expressed.' Tombs noted that the load factor of the twelve large – over 500 megawatt – pressurized-water reactors then in service worldwide was only some 60 per cent, compared to 80 per cent for Magnox.

I think in buying light water reactors one is buying on the basis of expressed confidence, and the fact that there would be a larger financial commitment to the solution of problems, but problems there certainly are – some of them quite difficult – both on the availability front, and perhaps in this country or any country also on the safety front.

Tombs left no doubt that for his part he favoured the AEA's steam-generating heavy-water reactor.

Five days later, on 18 December 1973, CEGB chairman Arthur Hawkins and his safety chief Roy Matthews took a stand that flatly contradicted their colleagues from Scotland. Challenged to justify the startling change in the CEGB position from the line he had taken only sixteen months earlier, Hawkins was defiant, indeed truculent.

I believe that I may have failed to convey to the Committee in the evidence which I gave in August 1972 that we had no need at that moment to order a new nuclear station. We had in mind that we needed to go on working at it with a view to resolving the problems about what was the best system to order when the time came to order. The time has now come to order.

Examination of the transcripts of the 1972 hearings offers little support for Hawkins's attempt to rewrite his earlier testimony.

Hawkins declared that there had been no significant change in the load estimates. He further declared that the decision to embark on this new plan had been reached before the October Yom Kippur war and the oil price increase. What, then, had changed, and why the urgency? asked the committee, especially in view of the SSEB opinion that 'there is no need for panic'. 'May I assure you,' responded Hawkins, 'that there is nothing in anything we are doing or suggesting at the moment which suggests there is a need for "panic". I prefer to say "crash programme".'

Just how 'crash' this new programme was to be Hawkins then revealed, for the first time in public. 'We would like to order in 1974 two stations . . . in 1975 one station, another in 1976, another in 1977, two in 1978 and two in 1979 – nine new stations – and nine more from 1980 and [sic] 1983 . . . With possibly two exceptions these are twin-reactor stations on the basis of 1200 megawatts to 1300 megawatts per reactor.'

The committee's collective jaw dropped. Members tried to elicit some explanation of the dramatic change in Hawkins's views since his

August 1972 testimony; but Hawkins insisted loftily that his views had not changed, and that the committee had misunderstood him sixteen months earlier. The committee, still somewhat in shock, then moved on to the question of choice of reactor. Hawkins did not mince words. 'We have, in effect, at least three distinct designs of AGR; they are all prototypes, and we are trying to rely on them as commercial reactors. This is a catastrophe we must not repeat.' Hawkins's views on the SGHWR had not changed. On 2 August 1972 he had dismissed the SGHWR out of hand: 'The SGHWR is . . . already out of date as a technology . . . a convenient stopgap if we needed one . . . we do not need one.' In December 1983, when Palmer reminded him of these 'very strong words', Hawkins replied bluntly 'I do not recall your challenging them last time'. Palmer insisted: 'I challenge them now.' Hawkins softened his tone slightly, and quoted back the committee's comment in its June report, which he called 'admirably worded', calling for a 'serious reappraisal' of the SGHWR: 'it might well be that there is little point in continuing it, sad as is such a conclusion after all the hopes of earlier years'. Hawkins said, with finality, 'We thoroughly agree with that.'

He then put in a word for the high-temperature reactor – provided that the government put up the money for a demonstration plant. However, he asserted: 'Since we need megawatts and nuclear units in the intervening period, we need to order what I have chosen repeatedly to call some bread and butter nuclear plant. We want something which is as proven as it can possibly be.' The 'bread and butter nuclear plant . . . as proven as it can be', turned out to be 1300-megawatt PWRs. Airey Neave, a much respected Conservative committee member, forthwith demonstrated the meticulous style of cross-examination that had earned him acclaim as an official prosecutor during the Nürnberg war crimes trial in 1946. To extract isolated quotations from Neave's cross-examination does scant justice to its tight and relentless coherence. It left Hawkins blustering and floundering.

Contrasting, as he saw it, the development status of the SGHWR with that of the PWR, Hawkins asserted: 'there is operating at the moment a PWR of the size we require in operation'.

NEAVE: A 1300-megawatt PWR?
HAWKINS: It is of the same reactor size as at Zion [Illinois] that we would be

ordering. It is limited because the largest turbine they could make was something between 1000 and 1100 megawatts. So it is something just under 1100 megawatts, and it is operating.
NEAVE: But you want 1300 megawatts, do you not?
HAWKINS: Look, the same sized reactor as at Zion, and I do not mind whether it is 1200 megawatts or 1300 megawatts. I am not going to quarrel about that.
NEAVE: I am going to quarrel about it because there is no 1300-megawatt light water reactor in operation in the world.
HAWKINS: All right. I said there was a reactor of the size we would order in operation in the world.
NEAVE: Did you not tell the Committee that you want to order a 1300-megawatt reactor?
HAWKINS: We would propose to go somewhere between 1200 and 1300 megawatts in reactor size.
NEAVE: I am suggesting to you that reactors of that size are not yet in operation and that the industry of this country has no experience of such reactors. I am also suggesting that your ordering of them at this time would cause delay.
HAWKINS: Well, we do not agree.
NEAVE: You do not agree?
HAWKINS: No.

The cross-examination continued in this acrimonious vein virtually throughout the session. Allowing for a longueur here and there, the entire session was an engrossing – not to say alarming – illustration of how elusive the CEGB could be, and yet how overweeningly sure of itself. How little foundation there was for this boundless self-assurance was revealed immediately the following day, when the Chief Inspector of Nuclear Installations, Eric Williams, once again appeared before the committee.

The central thesis of the evidence given by both Sir Arnold Weinstock and Arthur Hawkins had been the urgent need to construct nuclear stations to meet anticipated growth in electricity demand. Only PWRs, they insisted, could be constructed rapidly enough to meet this need. They waved aside all doubts about the actual track record of PWR construction and operation elsewhere, and refused to countenance any suggestion that the British nuclear industry might encounter difficulties in meeting the tight schedules and budgets their proposals implied.

One such difficulty, however, was fired across their bows the day after Hawkins's testimony. Eric Williams reminded the committee of his evidence nine months earlier: he had received no formal application for approval of a detailed PWR design. The CEGB had, it is true, filed an outline application for its Sizewell site, for the construction of a new nuclear station that might be based on AGRs, HTRs, PWRs or heavy-water reactors; and there was also the more recent Orford Ness application. But that was only the first step. From the time the Inspectorate received the actual detailed design, full approval for a PWR would still take some two years.

This two-year hiatus carved a deep gouge out of the confident timings and prognoses advanced by Weinstock and Hawkins. Yet it cannot have come as a surprise to them; it was after all merely a reiteration of the position delineated by Williams nine months earlier. Their casual dismissal of Williams's warning meant either an indefensible disregard for the essential legal requirements of the British nuclear licensing procedure, or an expectation that political manoeuvres would outflank the Inspectorate.

Nor was Williams the only stumbling-block to surface before the committee. Dr Larry Rotherham had been a member of the board of the CEGB for twelve years, and before that a senior staff member of the Atomic Energy Authority. Rotherham was a director of Fairey Engineering, which had been one of the members of the ill-starred consortium Atomic Power Constructions; his inside knowledge of the Dungeness B débâcle helped to give him a less than rosy view of nuclear forecasting. In a written submission to the committee just after Christmas 1973 Rotherham did not waste words on niceties. His memorandum got down to business in its first two sentences:

> The arguments for any one nuclear system compared with any other have gone on now for nearly twenty years and have been bedevilled by pseudo-economic arguments, differing ground rules, emotional and political judgments. It is virtually impossible to advance quantitative reasons for any conclusion to be generally acceptable, since any set of calculations can be matched by a comparable set leading to a different conclusion.

Rotherham spoke up for the merits of the British heavy-water reactor. He reminded the committee that in 1969 CEGB chairman Sir Stanley Brown had written them to say that 'We believe the SGHWR would be satisfactory for use on the Board's system'.

Rotherham listed the reactor's advantages, not least that of safety. Its fuel was enclosed not in a massive welded steel pressure vessel but rather in many separate pressure tubes, so that there was much less possibility of catastrophic rupture. He also pointed to the reliable performance of the 100-megawatt SGHWR prototype at Winfrith, as evidence that the design was proven in service and worth more consideration than the CEGB was giving it.

A fortnight later the committee received another message with the same tenor. It had invited comment from the doyen of British nuclear engineers, Lord Hinton; his response expressed a similar view. Modestly disclaiming any special insights into recent nuclear developments, Hinton nevertheless sounded a warning.

> In particular the Committee should remember that, for the last nine years, I have had access to little more than published information and to informal statements and that these (when they come from manufacturers and utilities) give the impression of being cautious and guarded; they leave me in doubt as to whether all the problems are revealed.

Hinton then reviewed the attributes of the various competing reactor types. His comment on light-water reactors was uncompromising:

> Light water reactors are not economical burners of uranium and their efficiency is so low that they cause thermal pollution. But the important question is whether, in our crowded island, they are safe. Many Americans are doubtful about their safety. The effectiveness of the emergency cooling arrangements is questioned. Light water reactors use very large welded vessels with many welded branches. I am assured that the technique of welding is now so advanced that these vessels can be considered as absolutely safe. But it is not so many years since a conventional boiler drum in the UK broke in half at the circumferential weld while it was being lifted into position and only three years ago the Generating Board attributed outage of many of its modern high pressure boilers to defective stub welding of branch pipes. It seems to me that in the last ten years the size and rating of the light water reactors has been pushed forward so daringly as to involve the possibility of hazard. All plants (even conventional plants) involve some measure of risk but it seems to me that of all the nuclear plants at present on the market the ones whose safety should be most strongly questioned are the light water reactors.

Hinton too came down on the side of the pressure-tube heavy-water reactors – the CANDU and the SGHWR: 'This is the gamble that I would take if (without further information) I had to make an immediate decision.'

By this time matters were moving swiftly on every energy front except the nuclear. On 8 January 1974 Prime Minister Edward Heath announced a shake-up in the structure of the government, with the creation of a new Department of Energy carved out of the old Department of Trade and Industry. The new Department had more than a passing resemblance to the one-time Ministry of Fuel and Power; but 'energy' was now the okay word, and 'fuel and power' had a distinctly fusty ring, so 'Energy' it was. Energy also got a new Secretary of State in Lord Carrington. He sat down to a bulging in-tray. The actions of OPEC were still reverberating through the international oil scene. Britain itself was gearing up to become an oil producer, amid furious debate about the ground rules for on-shore and off shore development, taxation and a myriad other details.

To Britons themselves, however, the energy issue of immediate concern was the work-to-rule by British coal miners. It had so severely eroded power station coal stocks that in mid-January 1974 the government had to resort to Draconian measures. Industry was ordered to cut back production to a three-day week; and the country began to suffer sequential power cuts in an effort to stretch the remaining reserves of power station coal. The official catch-phrase was 'SOS': 'Switch Off Something'. The CEGB had enough to worry about without further confrontation over its nuclear plans; indeed many people, especially within the Conservative government, devoutly wished that the CEGB had more nuclear stations and fewer coal stations, to reduce the leverage of the miners. But even among these nuclear supporters a good many believed that, amid the switching-off, the CEGB ought to switch off its plans to import PWRs.

Nevertheless, the PWR faction was soon on the attack again. Lord Aldington, chairman of the National Nuclear Corporation, and an influential senior member of the governing Conservative Party, took the stand before the Select Committee on 17 January 1974. Aldington told the committee that his new corporation was in the final stages of negotiations with the old consortia, TNPG and BNDC, to take over the responsibility for completing the AGR stations. Airey Neave, for the committee, asked 'whether the NNC is going to concentrate on the completion of these AGR stations before they turn over to a new technology in which our industry has no expertise'. Aldington responded, 'You will forgive me if I say that that is a slightly loaded question'; to which Neave replied 'It is meant to be'.

Challenged by Ron Brown to say whether or not the NNC was 'campaigning to ensure that British technology is promoted by the NNC', Aldington took issue with Brown's reference to a 'vast lobby for the American LWR': 'I do not view this matter as a matter of lobbies. I have not noticed a vast lobby in favour of the LWR; I have noticed a vast and very vocal lobby against it, but perhaps I have not looked in the same places as you have.' Aldington insisted that 'the business of the NNC is to reach the right decision based on the facts, and part of the facts is the reason why one wants the nuclear reactor'. The reason was 'to produce electricity in substantial amounts in the early 1980s . . . From all I am told by the CEGB who are responsible for this, it looks as if the cheapest way for Britain to get power in the 1980s is by means of the light water reactor.'

He defended this position stoutly against a steady barrage of sceptical questions from the committee. Yes, the British nuclear industry could build a Westinghouse PWR more quickly in Britain than it had been able to build British-designed AGRs. Yes, the Zion PWR in the US was a suitable precursor for a proposed '1200 to 1300 megawatt' British PWR, even though Zion was producing less than 1100 megawatts. No, he did not accept that there were no PWRs operating at more than 800 megawatts; he was being advised on that. 'It is a question of what will be operating, about which there will be a great deal of experience by the time we will be building and commissioning.'

There was, said Aldington repeatedly, a 'chicken and egg problem', about both safety approval and costs. Until the NNC issued a letter of intent to purchase a particular reactor, it could not obtain engineering details necessary to convince the Nuclear Inspectorate of its safety. On the other hand, unless it could obtain the Inspectorate's blessing, it could not with confidence issue a letter of intent. A similar vicious circle applied to cost-estimates. Without an order, it could not estimate costs; and without a cost-estimate, it was difficult to obtain an order.

Neave put his view bluntly. 'Do you not think it extraordinary that the CEGB should announce to this Sub-Committee a programme involving thirty-six large American reactors with no service experience without giving any details of the cost?' Aldington demurred.

> I do not think it is extraordinary that the CEGB should quite openly state to you what their present plans are, well knowing that their present plans are all subject to a major government decision . . . I do not think, with respect, you should criticize them just because they are unable to dot all the 'i's and cross all the 't's. They cannot do that until decisions have been taken.

Neither the scale of the proposed programme nor the reactor on which it was to be based gave Aldington a moment's pause. He bridled at Ron Brown's suggestion that he was inadequately informed about the track record of US nuclear plants; the exchanges were as heated as the sombre and intimate setting of the packed committee room could well accommodate.

Pressed on the role and activities of the Nuclear Power Advisory Board, Aldington would say only that

> It is an advisory body about which the members who join it and give advice do not talk outside . . . The number of times it meets is really a matter for the chairman of the board to announce if anybody really wants the information. On the first occasion we did announce that we had met. Also we announced that we had met on the second occasion, and we have met on several occasions since then.

He would not even confirm Neave's suggestion that the board had met the preceding Monday. All in all it made the Advisory Board sound more like a Masonic lodge. As events later proved, the Advisory Board had a great deal to be secretive about, at least for purposes of saving face.

On 22 January 1974 the Select Committee received a memorandum from the chief scientific adviser to the Cabinet, Sir Alan Cottrell. He was a metallurgist of international standing, and accordingly an expert in one particular discipline of special significance to the PWR: the integrity or otherwise of its massive steel pressure vessel. Cottrell gave the committee a short course in fracture mechanics, and concluded:

> 1. Rapid fracture, from large cracks or defects in thick sections, is in principle possible in steel pressure vessels under operational conditions. 2. In LWR vessels the estimated critical crack size for unstable growth is smaller than the wall thickness, so that the 'leak-before-break' safety feature is unavailable.

3. In these circumstances, the security of an LWR vessel against fracture depends on the maintenance of rigorous manufacturing and quality control standards; and on thorough, effective and regularly repeated examination of the vessel by the ultrasonic crack-detection technique. 4. The possible gradual growth of small cracks in highly stressed regions, by ageing and corrosion effects during service, needs further scientific investigation; as does also the effect of thermal shock from emergency cooling water in a loss-of-coolant incident.

It was an analysis that was to hang over the PWR for a long time to come.

The following day, 23 January, Sir John Hill, chairman of the AEA, appearing before the committee in person, differed comprehensively, if fuzzily, with Lord Aldington and Aldington's colleagues. Hill attempted repeatedly to retreat into expansive generalities about the essential role of nuclear power in the energy mix. But the MPs did not let him leave it at that. The result was an assortment of startling internal contradictions in Hill's testimony. Hill interpreted the CEGB's position in a surprising way. He said that he understood that they might need a 'fill-in' reactor while making the transition to the high-temperature reactor and eventually to the fast breeder. He also insisted that no one ought to assume the need to plan an ordering programme out to ten years hence; two or three years were plenty, and decisions on later orders could be taken when necessary. This was difficult, not to say impossible, to reconcile with the picture presented by Hawkins. Hill nevertheless insisted that 'I do not think that the difference [between his views and Hawkins's] is perhaps as great as you have made out.'

At one point Hill asserted that 'a steady programme of constructing two or three large nuclear reactors a year, really, to provide this base load of nuclear generation, is needed'. He then, however, agreed that there would have to be a gap of two years after the first order 'until experience is accumulated'. He also conceded that despite his avowed acknowledgement of the force of the CEGB's arguments for its plans, 'within 3½ years of placing the order for the first station we shall have at most two stations actually being built'.

Hill stubbornly defended the reactors of British design. He declared: 'I think we ought to restrict our programme to one thermal reactor of British design or evolution and to the fast reactor.' He devoted considerable time to extolling the virtues of the gas-cooled lineage; he was, however, reluctant to pick up the committee's frequent cues to elicit advocacy of the British heavy-water reactor. On the question of safety he insisted that 'All reactor systems have their own problems', and refused to be drawn into criticism of the light-water design. Hill fell back continually on the role and responsibility of the Nuclear Inspectorate. The chief inspector's would be the relevant opinion; he it was who would have to be satisfied. Hill's evidence, as the last witness to appear, gave a final stir to a pot already impenetrably murky.

By this time all the arguments from every faction had been amply ventilated: and were accordingly seen to be full of holes. A bare week after Hill's appearance the committee's report was complete. Published on 2 February, it left no one in any doubt about the committee's attitude to the issue, and to the evidence it had gathered. The text of the report was less than four pages long; but almost every line bristled with hostility to the CEGB and to its case for the PWR.

The CEGB were evidently mistaken in the views on nuclear capacity in relation to load growth which they put before us in August 1972. We need much greater assurance that their new plans are based on more valid assumptions . . . Mr Hawkins seemed to suggest that the main virtue of the PWR was that it was a thoroughly proven system which could be made available quickly. However, Mr Williams, the Chief Nuclear Inspector, told us that it would take him about two years to form an opinion on the safety of the PWR in British conditions.

The committee was clearly impressed by the strictures of chief scientist Sir Alan Cottrell, about the possibility of fracture of a PWR pressure vessel; the report quoted a key Cottrell paragraph at length. It was much less impressed with the economic arguments put forward in support of the PWR. 'No part of the evidence which we have received on capital costs is directly comparable with any other part; this leads us to suspect that, unless there is a great deal of operating experience with the system in question, no one can guarantee that any given reactor system is going to prove cheaper than another under actual operating conditions.' The last four words were in italics.

The committee's conclusion was blunt. 'We still strongly support the installation of new nuclear capacity, but no proposal to build

American light water reactors under licence in the UK should be approved by the Government on the basis of the evidence at present publicly available.' The committee repeated its warning about PWR safety, and closed by noting 'the enthusiasm of the South of Scotland Electricity Board for the SGHWR' – the British heavy-water design so long the darling of the committee.

Meanwhile, back in the real world, the Heath government had other problems on its collective mind. The three-day week was still in force, and candle-lit dinners were losing their romantic aura. Within a few days the coal miners had decided to turn their protracted go-slow into an all-out strike. On 7 February, demanding 'Who rules the country – the government or the miners?', Prime Minister Heath called a snap election for 28 February. To his baffled dismay he lost it. In so doing he also lost his credibility with the Conservative Party, a consequence that was to have longer-term implications even for British nuclear policy. The immediate impact of the election, however, was to bring Harold Wilson and Labour back into office – a party whose fondness for GEC and Westinghouse was distinctly muted, and whose flag-waving for British technology was unstinting.

When the Wilson Labour government took over from Heath's Conservatives the prospects for the Westinghouse PWR at once nose-dived. In the new Cabinet the Energy portfolio was taken over by Eric Varley, an ex-miner. In short order the miners' strike was settled – essentially by giving in to their demands. In an interview published in *The Times* shortly after his appointment, Varley promised that before any decision was taken about future nuclear plans there would be a debate in the House of Commons. That in itself was something of an innovation. It might not in truth greatly influence the eventual government decision; but it was rare for a government even to concede the opportunity to discuss nuclear matters in the Westminster forum.

The change of government also enhanced the influence of the trades unions, some of which, like the Institute of Professional Civil Servants, had already registered their opposition to importing American reactors. The reactor manufacturers were not, however, overtly crestfallen. Westinghouse and US General Electric were both placing full-page advertisements in the national press, extolling their nuclear technology. Atomic Energy of Canada Ltd, with a smaller promotional budget, made do with two-thirds of a page. It was frankly disconcerting to see newspaper ads for nuclear power stations – one of the more bizarre manifestations of the newly public aspect of nuclear controversy in Britain.

Meanwhile, Sir Arnold Weinstock too was talking to the press. It emerged that plans were taking shape for an international collaborative link between the National Nuclear Corporation, Westinghouse, and Westinghouse's French licensee Framatome. According to Weinstock, this three-way collaboration would make a powerful impact on the nuclear power plant export market, in which Britain's role had since the end of the 1950s been effectively non-existent. It was indeed a tempting notion – if it came off. Some commentators remained doubtful.

On 19 March the Royal Commission on Environmental Pollution announced that its next subject for study would be 'nuclear power and the environment'. The committee was chaired by Sir Brian (later Lord) Flowers, FRS, himself an eminent nuclear physicist, and a part-time member of the Atomic Energy Authority. His status in the nuclear establishment attracted sceptical comment: just how independent would this study actually be? As it was to prove, the Royal Commission was in no one's hip pocket.

On 29 March figures were published showing that the cost of the AGRs to date was already some £900 million higher than the original estimated total of £625 million. In April it was revealed that the CEGB had attempted in January to issue a letter of intent to Westinghouse, which would have pre-empted the government's decision. But the attempt was foiled by the snap election and its outcome. The CEGB was applying strenuous pressure on Varley for an early decision; but Varley refused to be hurried. Other pressures were also of concern: a delegation led by Walter Marshall, deputy chairman of the AEA, had departed at the beginning of March for the US, to study the work there underway on the integrity of PWR pressure vessels. On 27 June, Marshall was also to be appointed Chief Scientist in the Department of Energy, on the retirement of Sir Alan Cottrell. Marshall's dual role, at the AEA and the Department of Energy, was to attract stiff criticism, implying as it did an unambiguous conflict of interest, and an unwarranted reinforcement of the nuclear lobby in Whitehall.

By May fresh rumours had begun to surface: the Labour government was going to turn down the PWR proposal, and opt instead for the dark horse – the British steam-generating heavy-water reactor. The Department of Energy denied that any decision had yet been taken; and in the promised nuclear debate, on 2 May, Varley reiterated this denial to the House of Commons. Other speakers in the debate returned variously to their favourite themes, and the sound of grinding of axes echoed in Parliament Square. But the debate otherwise added little if anything to the arguments already exhaustively rehearsed throughout the preceding months.

One topic, however, already rehearsed but far from exhausted, received a sharp nudge on 7 June. Sir Alan Cottrell, retired as Chief Scientist but by no means put out to grass, wrote a crisp letter to the *Financial Times*, declaring himself profoundly uneasy about the safety of PWR pressure vessels. However strong his sentiments in his January memo to the Select Committee, his letter to the *FT* was even blunter; and it reached a substantially wider audience. If the government had still been tilting towards the PWR – which rumour by this time gravely doubted – Cottrell's letter would certainly have made the choice politically virtually impossible.

Rumour was correct. On 10 July, several weeks later even than his own expectations, Varley at last made the long-awaited announcement: the next nuclear programme would be based on the steam-generating heavy-water reactor. Lo and behold: the ugly British duckling was a swan. It was, however, not so large a swan, nor would its offspring be so numerous. The new programme would consist, in fact, of only 4000 megawatts of plant, rather than the 41,000 megawatts postulated in the CEGB proposal. This meant, more specifically, a total of six SGHWRs, each of 660 megawatts. Four would be built by the CEGB, and two by the South of Scotland Board. The National Nuclear Corporation would be invited to get to work at once on scaling-up the 100-megawatt design from the AEA's Winfrith prototype to the required commercial size. This time – so the story went – the scale-up would be a reasonable one, unlike the twentyfold scale-up attempted from the little Windscale AGR. It was a good story; but it still proved to be fiction.

Varley's announcement was greeted with joy by the heavy-water brigade. GEC and Sir Arnold Weinstock, however, were furious, and made no effort to conceal their fury. At the CEGB, Hawkins's deputy Donald Clark at once tendered his resignation. The PWR proposal had been Clark's particular baby; when it was thrown out he washed his hands of the whole matter. Varley had been careful, in his statement, to insist that the decision in no way reflected concern about the quality or safety of the PWR. The Nuclear Inspectorate were to continue and complete their generic assessment of the PWR, to keep the way open for it, should it in the longer term become more interesting. The PWR proponents in Britain were unappeased by this concession. In their view the decision to reject the PWR was indefensible; no token study could compensate for what they considered an opportunity squandered. They were, however, far from vanquished. It was, to be sure, the end of Round Two in the battle between the AGRs and the PWRs. But the PWR proponents did not abandon the field. They just retired to their corner, to await the bell for Round Three.

3 The unpronounceable reactor

Consider the following postulate. You are awaiting a crucial government decision, expected within the month, as to which reactor will be chosen as the basis of a new nuclear programme. You are also seeking approval to build a new nuclear station at a particular site, called Torness, on the Scottish coast southeast of Edinburgh. Do you wait until the reactor choice is announced and submit your application accordingly? Not if you are the South of Scotland Electricity Board, you don't. Instead, you submit an application for permission to build either AGRs, or HTRs, or PWRs, or BWRs, or SGHWRs.

In the circumstances you cannot be expected to go into detail about the design, or defend any detailed criticism of its economics or safety; the whole proposal is after all at a very early stage. Never mind that this will be the only opportunity for the public to register any objection to the proposal, the only opportunity to press the SSEB for answers to questions about its nuclear policy before that policy is implemented in steel and concrete. So much the better to get the irksome business of public participation out of the way while it can offer the least impediment to nuclear aspirations. The government shares your attitude; and the Torness inquiry duly convenes in mid-June 1974, to assess the SSEB proposal to build reactors of unspecified size and design, at an unspecified time in the future.

The Torness inquiry opened on 18 June 1974, in the small town of Dunbar. It was a strange charade. All the participants – the SSEB not least – knew that an official announcement on the choice of reactor was imminent. Yet the applicants firmly declined to respond to cross-examination about the characteristics of the plant they might build; that had still to be determined by the government. As arid exercises go the 1974 Torness inquiry must rank with the most sterile. Nevertheless, the 'Reporter', as the inquiry chairman was called, in due course found in favour of the 'application', such as it was: by which time the choice had fallen on the SGHWR. In the ensuing months and years, as the 1974 inquiry receded in memory, it was invoked repeatedly by nuclear officialdom to bless a succession of policy notions eventually having essentially no relationship whatever to the transactions of the original 1974 inquiry. As a demonstration of the public's perceived role in nuclear policy the fortnight of the Torness inquiry would be difficult to beat.

The government accolade conferred on the steam-generating heavy-water reactor did not long go unchallenged. Commentators pointed out a feature, hitherto little remarked, that could not be ignored if the SGHWR were to become a fixture on the British nuclear scene. It was a heavy-water reactor; ergo, it needed heavy water. Heavy water could, of course, be imported, probably from Canada. But if it were to be manufactured in Britain – an obvious corollary of any significant construction programme – the clear-cut safety hazard of a heavy-water plant must be recognized. Heavy water was manufactured by a process that involved the use of a very large inventory of hydrogen sulphide. Well known to school-children as 'rotten egg' gas, hydrogen sulphide is in fact frighteningly toxic. A concentration of only ten parts per million in air can cause sickness; 500 parts per million is lethal. Furthermore, at dangerous concentrations it overwhelms the sense of smell, and cannot be detected by the nose. Heavy-water plant staff work in pairs, with preassigned intervals between them, lest one of the pair abruptly keel over. An accident at a heavy-water plant might cause devastation to the surrounding community. Suddenly the safety of the pressure-tube reactor design was undercut by the hazard of its key service facility.

A more mundane but curiously annoying problem also hung over the steam-generating heavy-water reactor: its name. Other reactors with cumbersome names had the saving virtue of euphonious acronyms available – AGR, PWR and so on. Not so, unfortunately, the British heavy-water design: 'SGHWR' was even more unpronounceable than its full name. Even its most committed supporters paused before referring to it explicitly, undecided which of the two clumsy designations to employ. One commentator tried to encourage the term 'steamer'; one at length floated the idea of relabelling it the 'BTR', for 'British Tube Reactor'; but the coinage never caught on.

The final solution to the problem, when at last it came, was unhappily drastic.

For the nonce, however, the SGHWR was the chosen design of reactor for the next generation of British nuclear stations. Like it (the SSEB) or not (the CEGB), the electronuclear industry set to work to make it a functional reality. Earlier in the 1970s, indeed, the SSEB had got as far as suggesting that it might build an SGHWR station at Stake Ness on the Moray Firth; the idea came to nothing, but the SSEB now accepted the thought of an SGHWR station with genuine satisfaction. The CEGB did so with gritted teeth, conscious that it had suffered an acutely humiliating slap in the face.

Work on the new nuclear programme got underway, however, against a background of perceptible unease about the fundamental premise on which it was based. In 1972 the electricity supply industry had been anticipating at least a 3½ per cent annual growth in demand for electricity. In the immediate wake of OPEC's oil price rise of late 1973 the expectations of the electricity suppliers rose; the annual report of the Electricity Council, published in mid-1974, went so far as to anticipate an annual growth of 6¼ per cent in electricity use in the next five years. As matters were to turn out, not only did such growth not materialize: electricity use in Britain actually decreased. The sudden rise in prices jolted people into a heightened awareness of the cost of fuel and electricity; the fourfold rise in the world price of oil caused a dislocating slow-down in economic activity everywhere, further reducing the demand for energy.

In Britain the consequences for the electricity industry were a surprisingly close re-run of its experience a decade earlier. In the mid-1960s, the Electricity Council had been expecting electricity demand to grow at some 7 per cent per year. To meet this anticipated demand the CEGB ordered not only the AGR stations but also a series of enormous conventional stations fired by oil and coal, and using turbo-alternators larger than any previously built in Britain. As mentioned in Chapter 2, these 500- and 660-megawatt turbo-alternator sets – forty-seven of them in all – proved to be a long-running technical headache. Not until well into the 1970s did the CEGB succeed in ironing out the bugs in these big sets. Site problems and other troubles grew so severe that the Labour government in July 1968 appointed a Committee of Inquiry to investigate 'Delays in

Commissioning CEGB Power Stations'. Its report was published in March 1969; but its findings did not help much. Ten years later the CEGB would still be struggling to finish power stations ordered in the 1960s. The delays would have had a catastrophic impact on British electricity users, but for one perversely fortunate fact. The electricity demand these stations had been built to meet did not materialize. Electricity use from the mid-1960s to 1974 grew at an average annual rate closer to 3 per cent than 7 per cent. The uncompleted, delayed power stations were not missed; their output would have been superfluous in any case. On the CEGB's estimates it would, to be sure, have been cheaper than the electricity from existing, operating stations. But the delays, in some cases of more than five years over original construction schedules, made all such cost-estimates utterly unreal.

In the months following the nuclear policy decision of July 1974, observers realized that they were seeing a second manifestation of the same kind. The CEGB's forecasts, put forward by Arthur Hawkins in December 1973 as the basis for a programme of more than thirty PWRs, were soon seen to be absurdly unrealistic. The Electricity Council forecast of a 6 per cent growth rate was even more far-fetched. Ere long the government was facing yet another crisis of nuclear planning, the direct converse of that which had apparently been arising in late 1973.

In the wake of the CEGB's vast proposed programme, commentators had seriously doubted whether British boiler-makers and heavy electrical plant manufacturers could possibly cope with having as many as six new nuclear stations all under construction simultaneously. By 1976 they were wondering if the same industry could survive at all without the orders it desperately needed. Those power plants – both nuclear and fossil-fuelled – that had been ordered in the 1960s were still under construction, slipping ever farther behind schedule. But their output was not missed, because electricity use by that time was even less than it had been in 1973. The mismatch between forecasts and eventual outcome, both of supply and of demand, meant no blackouts and only indirect embarrassment. The two wrongs made a right, of a sort – provided you overlooked the cost of unproductive capital tied up in the overdue plants.

Not surprisingly, the mythology of official 'energy forecasting' began to come under withering cross-fire from the sceptical sidelines. If it was liable to miss the target by such wide margins so regularly, what was the point of it? And what did it really have to do with planning new power stations and other facilities? Were not these decisions actually taken for quite other reasons, and the 'demand forecasts' devised to justify them? It was of course a cynical view; but it was based on abundant empirical evidence. In the follow-through to the SGHWR decision it was to receive powerful reinforcement.

On 13 September 1974 Energy Secretary Varley published the report he had received from the Nuclear Power Advisory Board. As one commentator sourly observed, it was a 'schizophrenic' document. The board had been in fact irreconcilably split between supporters of the PWR and those who could not be persuaded of its virtues, who insisted that the palm should go to the British SGHWR. Recall that the members of the NPAB had been appointed precisely because of their long and presumably profound experience of the nuclear industry and nuclear technology. Yet these 'wise men', on the basis of the same information, came individually to conclusions that were diametrically opposed to those of half of their colleagues. Indeed, the NPAB reflected in claustrophobic microcosm precisely the deep-seated split over beliefs – about reactor safety, nuclear industrial policy and the comparative merits and demerits of different reactors – that had bedevilled the debate in the world outside their secret conclaves.

Before the end of the year another schism was being rumoured. Following the government policy statement of August 1972 – so long ago – a design team at the Risley site of the old TNPG consortium had undertaken work on developing a detailed design for a commercial version of the SGHWR. This design was to be the basis of the orders anticipated after Varley's statement in July 1974. A team for Risley presented it to a meeting of the British Nuclear Forum in November 1974, to general approval: from all but the CEGB. The CEGB, it transpired, was lapsing into its old habits again. Its engineers at the Berkeley research laboratories were already busying themselves taking the Risley design of SGHWR to pieces and redesigning it. It was an ominous sign. Quoted comments from various parties put the likely timescale for a firm SGHWR order at least a year, possibly even fifteen months' away.

In April 1975, after what was reported to be relentless arm-twisting behind the scenes, Dr Norman Franklin was persuaded to leave his job as chief executive of British Nuclear Fuels Ltd to assume the same title at the newly-fledged Nuclear Power Company, the operating arm of the National Nuclear Corporation. It was to be, as Franklin had clearly surmised, a thankless task. Soon after Franklin moved into the hot seat at the NPC, Eric Varley was translated from Energy to Industry, and his place at Energy was taken by Tony Benn. Benn had been Minister of Technology in the Labour government of the late 1960s, during which time he had been responsible for some of the main nuclear decisions then taken – the later AGR orders in particular. By 1975, however, Benn's earlier enthusiasm for nuclear power as the epitome of technological 'white heat' had significantly cooled. His relationship with his Whitehall advisers on nuclear policy soon became more adversarial than cooperative.

By this time Sir Arnold Weinstock and GEC were in the throes of reducing their holding in the NNC to 32 per cent. They were, however, retaining their management contract. Some observers wondered how dedicated Weinstock and GEC would be to managing a corporation whose key business – designing and building the new SGHWRs – was so obviously anathema to them. Benn, for his part, made little attempt to disguise his distaste for the entire capitalist ethos of GEC, auguring a less than cordial dialogue between the nuclear management and its government overseer.

In August 1975 the Department of Energy published a slim A5 booklet reprinting the evidence the Department had given to the Royal Commission on Environmental Pollution. Its carefree superficiality was breath-taking. The booklet devoted a solitary page – eight sentences *in toto* – to 'UK energy demand and the prospects of meeting it'. Plucking figures out of the air it arrived at an 'energy gap' that would arise by 1990, demand that 'would need to be met by non-fossil fuels'. In a matter-of-fact final sentence it concluded: 'This gap would be equivalent, e.g., to some 20,000 to 45,000 MW of capacity, based on nuclear fission, in 1990, and anything up to twice or three times as much in the year 2000.' This off-hand conclusion could only be construed to mean that the Department of Energy considered it plausible to anticipate designing, building and commissioning, in only fifteen years, nuclear stations whose output would be equal to the entire existing load on the CEGB system – some sixty-five reactors of the largest size then authorized for construction in Britain; and following this feat by doubling or even tripling it in one further decade. As an illustration of the tenuous official grasp on practical nuclear reality in Britain, the Department's 1975 evidence plumbed unexpected depths of absurdity.

Even this document, however, acquired an aura of rationality when set beside the evidence the AEA gave to the Royal Commission a month later. The Authority based its analysis on what it called a 'reference programme' assuming that Britain would have in operation by the year 2000 a total nuclear capacity of – wait for it – 104,000 megawatts, of which no less than 33,000 megawatts would be fast breeder reactors. Sir Brian Flowers himself apparently took issue with this surreal suggestion, so much so that the AEA hastily pulled back, insisting that it was never intended to be anything so gross as a 'forecast', merely an upper limit for analytic purposes. The AEA staff had nevertheless clearly regarded it as an attainable goal, casting profound doubt on their competence, to say nothing of their judgement.

By the autumn of 1975 one reactor design in particular was finding the climate ever more unpropitious. During the controversy of the early 1970s the high-temperature gas-cooled reactor (HTR) had received dutiful praise from every quarter and every faction. It was agreed to be eminently safe, technically advanced, and with unrivalled potential for development. It could, for instance, deliver high-temperature heat for industrial processes like steel-making; it was the only reactor type with this capability. It did not, alas, have the backing of an influential faction in Whitehall. The Dragon experimental HTR was being operated by the AEA on its site at Winfrith, next to the prototype SGHWR. But the Dragon was an international project, originated and sponsored by the Organization for Economic Cooperation and Development (OECD), the club of rich Western nations. Dragon's role had never been clearly defined; as the years passed and expenditure mounted, with no apparent prospect for early commercialization, the supporting countries at length lost interest. Britain's contribution grew larger, while Dragon's future grew dimmer. By the end of 1975 Dragon's fire was dying. No one wanted to finance a new fuel charge; and six months later the project was abandoned. The decision did remove one competing design from the running in Britain; but British nuclear planners could always find new ways to make their lives more complicated.

In October 1975 Energy Secretary Tony Benn and his Scottish counterpart, Secretary for Scotland William Ross, gave government approval for investment in the two SGHWR stations, the four-reactor CEGB station at Sizewell B and the two-reactor SSEB station at Torness. Work on the stations could not, however, be started pending clearance of the detailed designs and issuance of nuclear site licences by the newly-constituted Health and Safety Executive, which had absorbed the Inspectorate of Nuclear Installations. Chief Nuclear Inspector Eric Williams retired in December 1975; his successor, Ron Gausden, took office determined to maintain the Inspectorate's stubborn if under-financed independence.

The nuclear industry, meanwhile, was not the only industry subjected to scrutiny with a view to reorganization. A committee of inquiry – chaired, as it happened, by Lord Plowden, a former chairman of the AEA – recommended in January 1976 that the CEGB, the Electricity Council and the twelve area boards be amalgamated into a single Central Electricity Board to serve England and Wales. This recommendation for yet more centralization of the electricity supply industry was not universally welcomed. The performance of the industry was by this time receiving sceptical scrutiny by various independent groups based in universities and environmental organizations. In general they doubted whether making the electricity suppliers even more institutionally powerful would necessarily reduce the incidence of arrogant incompetence thus far all too prevalent. Their doubts were shared by, among others, the Liberal Party; and the consequent policy struggle was in due course to have one unexpected result.

In early 1976 the government announced that GEC would indeed be reducing its shareholding in the NNC, to 30 per cent. Earlier speculation about the fate of the 20 per cent shed by GEC was answered in the least encouraging way for those who looked for a broad involvement of private industry in the reactor-building company. British Nuclear Associates, the hold-overs from the old consortia, were to remain with only 35 per cent; the AEA would pick up GEC's cast-off shares, bringing the AEA holding to 35 per cent. It was even reported that Energy Secretary Benn wanted the government, through the AEA, to acquire at least 50 per cent of the corporation.

For nuclear buffs, the main event of February 1976 took place on 5 February; in fact there were two events, within twelve hours of each other. At 5 P.M. the first reactor at Hinkley Point B was connected briefly to the electricity grid: the first AGR station to start up – in nuclear parlance, to 'go critical'. It was, to be sure, some four years behind schedule, but better late than never. Eleven hours later, at 4 A.M., the first reactor at Hunterston B in Scotland likewise went critical. The celebrations were brief; both reactors were shut down again after operating for only a few hours. Industry scuttlebutt had it that the teams at the two stations had been engaged in a race to see which could start up first. Events were to suggest that a better objective might have been to see which could operate longer and with less embarrassment.

In the official energy citadel of Thames House South on Millbank in London, Energy Secretary Tony Benn was already making his presence felt. At his behest the Department was drawing up an Energy Policy Review. When at length it was published in draft form, in February 1977, it was to mark a dramatic advance on the off-hand back-of-the-envelope evidence submitted to the Flowers commission. On 22 June 1976, in pursuit of his stated aim for greater public awareness of and participation in energy policy, Benn staged a one-day 'National Energy Conference' in London. Some fifty-six speakers, of varying eminence and responsibility, were each given a strictly rationed five minutes to make their views known. The gathering was a fascinating exercise in axe-grinding, making the political nature of energy policy abundantly evident as factions and interest groups defined and defended their turf.

One of several key documents newly to hand was the first published 'Corporate Plan' prepared by the CEGB. It bore scarcely any resemblance to the picture painted with a heavy hand by Arthur Hawkins thirty months earlier. Far from foreseeing electricity demand growing at nearly 5 per cent per year, it assumed a maximum growth rate of 3.4 per cent for at least five years, and a more probable rate of only 1.3 per cent. A corollary of this drastically reduced expectation was a similar reduction in the need for new generating plant. Where now were the thirty-plus PWRs so desperately demanded by the same people so recently? Why, indeed, did no one in official government authority trouble to ask? The text of the Corporate Plan now raised all manner of doubt about the ability of the British nuclear industry to manufacture even the modest amount of plant envisaged. In peevish tones the plan insisted sulkily that, had the PWR been approved, UK manufacturers would have invested in new production facilities. It did not, however, reconcile this assertion with the CEGB's own severely truncated requirements, as outlined in the same document.

The woes of the AGRs had been of course so long in train that few could be surprised by any fresh cock-up. Unhappily, however, by mid-1976 it was becoming apparent that all was not well with the SGHWRs either. The reference design for the proposed 660-megawatt units was completed in July 1976. As the more cynical had foretold, the full-scale SGHWR had turned out to be a great deal more expensive than originally expected. But this alone was not the crucial factor; nor, for that matter, was the by now obvious excess of generating capacity in the country. The CEGB, from having wanted, less than three years earlier, to order nine huge nuclear stations by 1978, was now having to shut down existing stations because they were not even being called upon from one year to the next. Nevertheless, the axe hovering over the SGHWR was being wielded not by the CEGB but by the government. As the economic status of Britain grew steadily bleaker, the government decreed drastic cuts in public spending; and one of the cuts was the £45 million that the CEGB had been preparing to spend on initial contracts for the Sizewell B SGHWR station. Accordingly, no order would be placed before 1978 at the earliest. The CEGB shed few tears at the thought; but the nuclear power station builders were distraught.

So, for that matter, was the Select Committee on Science and Technology. Its membership still numbered most of those who had staunchly pressed the case for the SGHWR in 1974. To see their darling once again in such straits – at the hands of the government that had so recently anointed it – was enough to make the committee break out in another rash of hearings. This time the hearings were held under the aegis of the General Purposes Sub-Committee, set up to see how past committee recommendations 'were working out in practice', as chairman Arthur Palmer, MP, explained on 2 August 1976 to the Sub-Committee's first witness, Energy Secretary Tony

Benn. He then added that press statements and comments in the House indicated that the SGHWR was in difficulties: hence the new hearings.

Palmer was at pains to stress that the committee had not actually 'recommended' the SGHWR; all it had said was, 'In this connection we note the enthusiasm of the South of Scotland Electricity Board for the SGHWR'. No subsequent unpleasant surprises about the cost and complexity of scaling-up could therefore by implication be laid at the doorstep of the committee. Be that as it might, the committee took a dim view of the latest developments, and expected the Minister to answer for them. This Benn did, in spades.

The rumours were quite correct; indeed they did not go far enough, nor did they convey the sensation that Benn forthwith disclosed to the committee. He revealed that no less than Sir John Hill, his own chief nuclear adviser, chairman of the AEA, had written to him a week earlier to recommend the abandonment of the SGHWR – the AEA's own reactor. Benn had asked Hill's permission to make the letter available to the committee, and indeed to the public, in line with Benn's desire for more informed public discussion of nuclear issues. Palmer read out Hill's conclusion: 'For a variety of reasons, the SGHWR programme looks less attractive than it seemed two years ago and, on balance, there is a consensus opinion (noting that SSEB dissent) that the programme be replaced by AGRs or PWRs.' Hill could scarcely have put the matter more bluntly. Asked for his reaction to this, Benn responded:

> The reasons given are lower electricity demand and public expenditure stringency. If that is the case with the SGHWR, it is bound to raise the question whether they are not really throwing doubt upon a thermal programme, at this moment anyway. The first point I made to them is that it does not follow at all that if you stop the SGHWR for the reasons given, you will be able to establish a case for another thermal system now.

Benn then rebutted all the subsequent conclusions in similarly robust vein.

The dissent recorded by the SSEB also came in for comment. SSEB chairman Frank Tombs had been, of course, the most outspoken advocate of the SGHWR during the controversy of 1973–4. His feeling by mid-1976 appeared to be that the SGHWR was being drastically over-designed: that the NNC, which had never wanted to build the reactor in the first place, was incorporating so much margin of safety that it pushed the system costs far beyond those necessary. Benn denied any suspicion that NNC was actively impeding the SGHWR by superfluous safety margins, or that this might be a malicious effort to nobble the SGHWR and clear the field for yet another foray on behalf of the PWR. The committee's minds were nevertheless set on seeking out any possible conspiracy: had Hill been forced into submitting this note by CEGB and NNC pressure on the AEA board? Benn declined to express an opinion on this point, but added:

What I said to [Hill] – and I think this is bound to be the case – was that if the AEA recommend the cancellation of their own system, it is bound to impact on the credibility, not only of British technology, but on the credibility of other systems that the AEA have sponsored. I quite specifically drew his attention to the fact that this could well have an effect upon the credibility of the AEA's case on the fast breeder reactor, on which, of course, they are very keen on an early decision.

The MPs pressed Benn to agree that a switch to light-water reactors would stir up renewed debate about nuclear power. Benn saw nothing wrong in that. 'I think that we are going to have a nuclear debate anyway . . . quite candidly, I do not think that anything but good can come of that.' It was a refreshingly different outlook from that which had dominated official nuclear opinion for more than two decades. At the end of the day, however, the overriding message was that within the nuclear sanctum the fists were flying yet again, and yet another policy was about to bite the dust.

One intriguing committee question was nevertheless ducked by Benn, albeit not without some suggestive hints. Committee member Kenneth Warren asked, with crisp directness and no preamble, 'Does Dr Walter Marshall agree with Sir John Hill's paper?' Benn hesitated, then inquired, 'May I ask for a ruling, chairman? Am I obliged to give my assessment of what the other participants think of each other's views? I think that I am on very dangerous ground.' The chairman left it to Benn; and Benn left it thus: 'I could be extremely interesting on this subject. I could go on for a very long time about what I am told by people about other people, but I think the Committee must rely on first-hand information.' It was a tantalizing glimpse of what would later become a first-class row, even by the rowdy standards of the British nuclear establishment.

While the AEA was counselling cancellation of the SGHWR, and the CEGB was sitting on its hands, perfectly happy to order no more stations of any kind indefinitely, Britain's power station manufacturers were slowly starving to death. Their plight had become so acute that the Central Policy Review Staff of the Cabinet Office had been directed to carry out an urgent study of the boiler and turbo-generator industries, to see what the government might do on their behalf. No station of any kind had been ordered since the Ince B oil-fired station in 1973. By autumn 1976 it was generally agreed that no SGHWR could be ordered before 1979 at the earliest. The Marshall report on the safety of PWR pressure vessels remained at the time unpublished; but the SSEB, flexing its muscles, was asserting its view that Marshall's findings did not significantly alter the case about the safety of PWRs in a British setting. The Scottish Board continued to insist that if PWRs were to be judged as stringently as the CEGB was judging the SGHWR, the PWR might well not measure up to the CEGB's safety criteria. The SSEB's outspokenness might have owed something to the fact that SSEB chairman Frank Tombs had been appointed to succeed Sir Peter Menzies as chairman of the Electricity Council. The switch would not take place until April 1977; but Tombs was wasting no time putting down his markers for the policy struggles to come.

With the SGHWR beginning to buckle at the knees, the PWR and AGR packs began to circle round it, waiting for an opening. Benn and the government were reserving their decision; but the failing health of the heavy plant manufacturers meant that a move of some kind could not be long delayed. Matters were further complicated for the nuclear planners by the publication of the sixth report by the Royal Commission on Environmental Pollution. At once christened the 'Flowers Report', after the commission's chairman, Sir Brian Flowers, the sixth report, entitled *Nuclear Power and the Environment*, became an instant classic. In lucid but magisterial terms it laid out its view of the issues enveloping nuclear power, with supporting arguments of impeccable authority. It rode the front pages and the leader columns for days, and was thereafter brandished by both proponents and opponents of nuclear proposals in every conceivable venue. Its most controversial findings challenged official British policy about plutonium, reprocessing and the fast breeder; we shall return to these topics in Parts II and III. But it also took discreet issue with the whole foundation of energy policy in the country.

The Flowers report even postulated an alternative energy strategy – one in which the longer-term use of electricity fell significantly short of that anticipated in official Department of Energy statements. The possibility brought little cheer to British fuel and electricity supply planners. It had of course been advanced by less august bodies since the early 1970s, usually in connection with environmental critiques of economic growth. By late 1976, the prospect of much-reduced growth in demand for fuel and electricity was amply plausible, even to the essentially establishment membership of the Royal Commission. It was not only plausible; it was happening. Since 1973 the use of electricity in Britain – confidently expected, by the CEGB and its adherents, to grow by nearly 5 per cent a year – had not only not grown at all, it had actually fallen. Official planners insisted that the decrease was a temporary phenomenon, a result of the world economic recession and the accompanying drop in industrial activity. Be that as it might, the decrease posed a major problem for British electricity planners – and a worse one for the manufacturers involved.

By December 1976 the British Nuclear Forum, a trade association of nuclear manufacturers, was urging the British government to decide on a reactor type and give the go-ahead for an order 'as soon as possible'. The Nuclear Power Company was carrying out a six-month review of reactor possibilities; NPC deputy chairman Jim Stewart reminded the Forum meeting that 'We have not had an order for a station in this industry since 1970. We don't want to press the government very much at this point, but we do want to see the colour of their eyes.'

Energy Secretary Tony Benn meanwhile let it be known that the Labour government did not, in its turn, much care for the colour of Arthur Hawkins's eyes. Despite the absurdity of his nuclear plans in 1973–4, Hawkins, as was customary, had in due course received his knighthood for services to the electricity supply industry. The Labour government, however, was not overly impressed with the services rendered; Hawkins's stormy tenure as CEGB chairman was terminated after a single five-year contract. His seat was to be filled, as of 30 June 1977, by Glyn England, moving up from chairing the South Western Electricity Board. England was to find that when the knighthoods were being handed out it was better to be a headlong proponent of nuclear plans, however far-fetched.

On 16 December 1976 the Central Policy Review Staff of the Cabinet Office – known as the 'think tank' – published its long-awaited report on *The Future of the United Kingdom Power Plant Manufacturing Industry*. It declared flatly that,

> Without appearing to overdramatize the situation, the position is that existing order books and the financial strength of some of the companies are not sufficient to enable the industry to survive in its present form. If the industry were required to undergo a major contraction in the next few years it would be unlikely to survive, either as an internationally competitive producer of power plant or even as a supplier of the full range of power plant required for the home market only.

The 'think tank' put forward five urgent recommendations for government action. Chief among them was 'a government commitment for a firm and steady programme of power station ordering over the long term, but starting now'. The report added that 'Without a firm commitment at the present to a nuclear system the industry has no hope of developing an exportable product for five years and probably ten'. More specifically, after castigating the CEGB's erratic ordering pattern of preceding years, the report urged the CEGB to give a firm contractual undertaking to order 2000 megawatts of new stations per year for ten years.

Outgoing CEGB chairman Sir Arthur Hawkins, true to form, dismissed this proposal: to build power stations that the country did not require would add significantly to everyone's electricity bills and further disrupt projections of electricity demand. In the light of Hawkins's own PWR proposal less than three years earlier, which would have launched a ten-year programme of orders on twice the scale of that put forward by the 'think tank', there was a certain irony about his reaction to the CPRS report. Hawkins left no doubt that should the government instruct the CEGB to order new stations sooner than it desired, the CEGB would expect government compensation – especially if the stations were coal-fired, like the long-delayed Drax B mentioned by the CPRS.

The CPRS confronted yet again the long-running wrangle about reorganizing the power plant manufacturing industry – this time fossil-fuel as well as nuclear – to match its potential market. It suggested that the existing boiler-makers should merge to form a single company, as should the existing turbine-manufacturers. As usual this dispassionate assessment was greeted by stubborn distaste on the part of the existing companies, between which little love was lost.

On 26 January 1977 the Select Committee on Science and Technology published its report on the SGHWR programme. It was a waspish, exasperated document.

To sum up: the present renewed debate on reactor choice revolves around three main issues: the cost of the different reactor systems; the safety of the different reactor systems; and the effect of the choice of reactor system on the future of the domestic nuclear and non-nuclear design and manufacturing industry. It is a sad reflection on our decision-making machinery, and on the quality of expert advice given to successive governments, that, seven years after the last nuclear station was ordered, and after extensive private and public debate, sufficient information is apparently still not available on any of these points for the country to proceed with confidence – at whatever pace – to the construction of new nuclear power stations . . . Although two and a half years have elapsed since the adoption of the SGHWR system for the next series of reactors, the reactor has neither been designed to agreed parameters nor accurately costed, and, in consequence, neither the opponents nor the supporters can argue their case with the ability to carry conviction in the minds of others.

It was a cry from the heart.

The committee reserved particular wrath for Sir Arthur Hawkins.

If Sir Arthur is in fact primarily concerned to secure financial support from the Government for the SGHWR, and to absolve his Board from financial responsibility for a decision which they have never liked, we believe it is unfortunate that he did not make that position clear from the start.

The committee recommended that work continue on the SGHWR until the NPC had completed its review of reactors, and that this review include comparative costings on a consistent basis with respect to operation and safety. The committee's patience was patently wearing thin; but it was to be strained yet further.

In April 1977 Frank Tombs took over the reins of the Electricity Council. His precursors in the council chair had tended to take a back seat in policy-making, leaving the current CEGB head to make the running. Such self-effacement, however, was not for Tombs. In accepting the chair Tombs apparently stipulated that he would be looking forward to an imminent reorganization of the electricity supply industry in England and Wales, along the lines suggested by the Plowden commission in January 1976. The commission had proposed that the Electricity Council, the CEGB and the twelve local area boards be amalgamated into a single central electricity authority; rumour had it that Tombs had been promised the control of this mega-body if in the interim he would be content just to chair the CEGB. As so often happened in the British electricity business, plans did not work out quite as expected – the plans of Tombs included.

In late June 1977, in a surprise move, boiler-makers Clarke Chapman and turbine manufacturers Reyrolle Parsons announced that they were to merge. The decision comprehensively derailed discussions about merging Clarke Chapman with Babcock & Wilcox, and Parsons with GEC, as proposed by the Central Policy Review Staff. The new company, in due course christened Northern Engineering Industries, was not what the independent analysts had suggested at all – another instance of industrial policy profoundly at variance with industrial practice. Its emergence meant that there would thenceforth be a continuing over-capacity for both boiler-making and turbine manufacturing for Britain's potential domestic market. The merger further aggravated the urgency of need for new power station orders, while complicating their allocation between the desperate supplier-companies.

In late June Energy Secretary Benn announced that the government was instructing the CEGB to place an order for the long-delayed second unit of the Drax coal-fired station in Yorkshire. The CEGB, already facing stiff criticism for its current over-capacity of generating plant – some 30 per cent above peak demand, with a great deal of further capacity already under construction – was deeply unhappy about the Drax B directive. It demanded compensation from the government for what it claimed would be a superfluous and premature investment, undertaken not to fulfil electricity supply requirements but to keep power station builders from collapse. As for the nuclear manufacturers, the Drax B directive was another body-blow. Granted that it set a precedent for ordering new plant ahead of necessity: it also made the necessity for new nuclear plant even more remote. The dispute was to go on simmering.

On 27 June Benn sacked his Chief Scientist, Dr Walter Marshall. The official announcement stated only that 'in view of the important decisions concerning nuclear policy that will need to be taken in the near future, and the significant role of the AEA in this area', Benn had asked Marshall to resume full-time work as deputy chairman of the Atomic Energy Authority 'as soon as possible'. Reports indicated that Benn and Marshall had parted company on the worst of terms, occasioned by Marshall's insistence on an early commitment to new orders for nuclear plant, and Benn's scepticism about the desirability or necessity of such orders, especially concerning the PWR. The prickly nature of their relationship was made disconcertingly clear in a brisk exchange of media salvoes. Marshall had declared that there was no such thing as a 'nuclear lobby' in Britain. Benn, asked to comment, observed tartly that not only was there a nuclear lobby, but 'Walter shaves part of it every morning'. Marshall's sacking reinforced the nuclear establishment's conviction that Benn was 'anti-nuclear', and made Marshall a martyr in nuclear eyes. The removal of Marshall infuriated the powerful core of nuclear supporters in Whitehall corridors – not least Benn's own Permanent Secretary, Sir Jack Rampton. In due course Marshall was to have remarkably sweet revenge.

In mid-July the National Nuclear Corporation delivered to Benn the report prepared by its operating arm, the Nuclear Power Company, on the prospects for the different reactor types. When in due course Benn published the report it proved to be yet another manifestation of the industry's chronic schizophrenia. It found that there could no longer be a credible case for building SGHWRs; but it then recommended a programme mixing AGRs and PWRs. As many commentators noted, this would guarantee the worst of both worlds, and multiply the already daunting problem of excess capacity, both on the electricity grid and in power station manufacturing. For its part the CEGB, under its new chairman Glyn England, was reported to be 'embarrassed' by this latest demonstration of the nuclear industry's inability to reconcile its perennial differences.

By the end of the summer the Nuclear Installations Inspectorate had completed its three-year study of the generic issues about the safety of PWRs. It proclaimed itself satisfied that the outstanding safety issues 'are not such as to prejudice an immediate decision in principle about the suitability of the pressurized-water reactor for commercial use in Britain'. Nevertheless, no such 'immediate decision' was forthcoming, even in principle.

Be that as it might, by November 1977 the CEGB had recommended to Energy Secretary Benn that a two-reactor AGR station be ordered soon, with a similar station in Scotland. In the CEGB's view the following order ought then to be for a PWR station. This would however require so much further design work that no PWR could be ordered before 1982. The National Nuclear Corporation, however, was reported to be pressing Benn to move immediately to PWRs. The AGR–PWR rivalry was by this time so intense that proponents of each type were openly itemizing the problems of the other, at least in the nuclear trade press. Hitherto a tacit gentlemen's agreement had kept all the various factions courteously insisting, in public at least, that all the various types were as safe as each other, and that any nuclear station was better than no nuclear station. Before the end of 1977 this tacit agreement had been intriguingly breached.

Neither faction was lacking ammunition. Since their nip-and-tuck start-ups in February 1976 both Hinkley Point B and Hunterston B had been gingerly raising power and carrying out tests. In June 1977 a water pipe at Hinkley Point B ruptured, cutting off the cooling water which kept the temperature of the concrete shielding below damaging levels. Site staff had to rig an impromptu arrangement of fire-hoses to protect the shielding. The incident, although embarrassing, was of only minor consequence for the station: not so, however, the accident which struck Hinkley's Scottish sister station three months later. This time jerry-rigged pipework was not the solution but the cause. In October 1977 staff at Hunterston B carried out a routine shut-down of the second reactor at the plant; but they overlooked something. Earlier water-supply problems had prompted the installation of a temporary pipe with a direct connection to the sea. Unfortunately, when the reactor was shut down, the drop in pressure inside its core reversed the direction of flow in this temporary pipe. As a result several thousand gallons of corrosive salt-laden raw seawater poured into the stainless-steel core of the reactor.

Only weeks before, the SSEB had leafleted its customers with a brochure proclaiming that the advent of AGR power on the system was going to mean lower electricity bills. The seawater influx, however, not only necessitated shutting down Hunterston B2 for more than a year; replacement of the complex steel insulation alone cost an estimated £14 million, and supply of replacement electricity from other stations added a further £50-odd million to the total bill. The SSEB, in a spectacular example of nuclear doublethink, eventually asserted that this bill for replacement electricity merely demonstrated the clear-cut advantages of nuclear power.

By late 1977, PWR supporters were supplying knocking copy to industry magazines drawing attention to the embarrassments at Dungeness B, Hinkley Point B and Hunterston B. In retaliation AGR supporters were noting the lengthening construction times and rocketing costs of PWRs in the US, and the accompanying catalogue of plant cancellations. To interested bystanders it was refreshingly forthright, and a welcome change from the mealy-mouthed good manners of nuclear industry contributions to the public colloquy as previously pursued. It did, however, prompt some commentators to note that both factions could well be correct: that neither the AGR nor the PWR looked like much of a bargain.

On 25 January 1978 Benn at last broke his prolonged silence. The SGHWR was to be abandoned – writing off its development costs of £145 million, a point that Benn's Commons statement did not happen to mention. The CEGB and the South of Scotland Electricity Board were each to order a new nuclear station as soon as possible – each one a twin-reactor AGR station. Benn's statement then went on to say this:

> The electricity supply industry have indicated that, to establish the PWR as a valid option, they wish to declare an intention that, provided design work is satisfactorily completed and all necessary Government and other consents and safety clearances have been obtained, they will order a PWR station. They do not consider that a start on site could be made before 1982. This

intention, which does not call for an immediate order or a letter of intent at the present time, is endorsed by the Government.

Pressed on every side to expand on this Delphic utterance, Benn asserted that the passage on the PWR had been 'very carefully worded to reflect the spirit of the agreement' between all parties to the official decision. To Opposition spokesman Tom King he added that 'On the PWR he will understand, if he reads the statement carefully – it was drafted with great precision – why I do not want to go beyond what I have said.' *The Times*'s report of the Commons debate noted that this reply was greeted with '(Conservative laughter)'.

In the fullness of time, the consequences of Benn's statement led many of those involved to wish that it only hurt when they laughed.

4 Sizewell – that ends well?

On 28 March 1979 the pressurized-water reactor in the second unit of the Three Mile Island nuclear power plant in Pennsylvania suffered the world's most serious civil nuclear accident. Nine months later, after fifteen years of havering, the British government, with its usual impeccable timing in nuclear matters, at last gave the long-sought unambiguous go-ahead to import American PWRs into Britain.

In the two years that separated Tony Benn's cagey statement of January 1978 and the official go-ahead for PWRs in December 1979, British nuclear power policy proceeded much as it had before – with one major difference. As we shall discuss in Part II, the general public had by this time become openly, actively and vociferously aware of nuclear power issues, in the wake of the controversy over the plans of British Nuclear Fuels Ltd to build a new 'reprocessing plant' at its Windscale site. The decision to order two new AGR stations, for the CEGB at its existing site at Heysham, in Lancashire, and for the SSEB at the virgin site of Torness, south-east of Edinburgh, provoked vigorous public protests, particularly at Torness – including demonstrations and occupation of the site by objectors, requiring the intervention of the police.

Otherwise it was business very much as usual – that is, confused and riven by internal conflicts. In June 1978 Lord Aldington, chairman of the National Nuclear Corporation, told a luncheon meeting of the British Nuclear Forum that there was still a need for a 'strong definition of the roles' of the NNC's Nuclear Power Company and the CEGB in nuclear design and construction. According to the trade press Aldington commented that 'having hired a dog to bark, the CEGB should ask itself what sort of barking it should do'. It was not clear at present what the NPC was supposed to do for the CEGB, over the whole field of design, procurement and supervision of construction – and what the CEGB intended to do itself. Someone was certainly barking up the wrong tree.

The 1978 up-date of the CEGB's corporate plan, published in June, introduced a new argument into the nuclear case. Previous proposals for nuclear plant orders had been based on estimates of future growth in electricity demand, and the consequent need to increase the total generating capacity on the supply system. In the June 1978 corporate plan, however, the CEGB declared that comparison of anticipated costs of power station types 'demonstrates that it could be economic to install new nuclear generating plant before it is required to meet increased demand for electricity'. From the nuclear point of view this was a convenient discovery. Since electricity demand was no longer growing, and since the available capacity on the system was already conspicuously surplus to requirements a different reason to order nuclear plant was obviously welcome: at least to those who wanted to order nuclear plant willy-nilly.

A corollary of this argument also surfaced, to the effect that in due course the existing generating plants would come to the end of their useful lives, and require replacement: and their replacement would have to be nuclear, since fossil fuels would by then be too expensive. To be sure, the assumptions underlying this new CEGB argument had still to be tested; and in due course they were going to be, with far from persuasive results. In the meantime, however, the CEGB and the NNC made vigorous use of the argument in briefings, lectures and lobbying.

On the other hand, the CEGB did not at once rush to place the expected order for its new Heysham B AGR station. One of the reasons for its hesitancy might have been the performance of the only two AGR stations by that time in operation. By March 1978 the four reactors at Hinkley Point B and Hunterston B had annual 'load factors' – actual output as a fraction of the maximum possible – from 32 per cent down to less than 22 per cent. Another reason for CEGB reluctance might have been the continuing troubles at Dungeness B, despite optimistic utterances from the Kent coast. In mid-1978 the Dungeness B project manager was quoted as believing 'that the engineering problems which have plagued the station have finally been solved, and some valuable lessons have been learned'. He expected the first reactor at Dungeness B to be 'producing power by 1980'. As it was to turn out, he was not even close.

A full year later, in March 1979, the industry magazine *Nuclear Engineering International* commented wearily:

The situation is still confused and hesitant. The design contracts for the two new AGRs at Heysham and Torness have still to be placed with the Nuclear Power Company, although work is proceeding on a day-to-day basis; the CEGB has still not decided what type of nuclear steam supply system [reactor and ancillaries] it will adopt for the first British PWR; [and] the reorganization of the National Nuclear Corporation and NPC has not been settled.

The Three Mile Island accident, on 28 March, did nothing whatever to dispel the fog. British nuclear scientists and engineers trooped with their fellows from other European countries to visit the stricken reactor; British nuclear authorities proclaimed smugly that of course British regulatory requirements and emergency plans were far more rigorous than those of the slap-dash ex-colonials. The CEGB stressed that Three Mile Island used PWRs from Babcock & Wilcox (no relation to the British firm whose name was then the same). These PWRs, it insisted were quite different from the Westinghouse PWR that was once again leading the field in British PWR planning. British AGR people permitted themselves a chorus of discreet 'I-told-you-so's'. They had to be discreet, not least because they were standing in a minefield of their own making. Onlookers noted drily that the AGRs were certainly impressively safe, in that the safest reactor was one that had never started up.

The AGR story took another twist in July 1979. By this time almost everyone with any standing in the matter had lamented the absence of any replicable standard design of AGR station, from Dungeness B onwards. Nevertheless, at the end of July the CEGB awarded a design contract to Parsons Engineering for the new Heysham B AGR station, for a 660-megawatt generating set with six exhausts; whereupon the SSEB chose GEC's rival four-exhaust design for Torness. So much for replication.

One expectation was, however, fulfilled: to no surprise whatever the CEGB asked the Nuclear Power Company to negotiate a licence to build a Westinghouse 1200-megawatt PWR in Britain. The CEGB stoutly denied that any specific PWR design had yet been chosen, despite the request; but few observers doubted that the choice would in due course be revealed to be the Westinghouse version. The CEGB let it be known that it would announce its choice and apply for the necessary consents the following year, with actual construction to commence in 1982. The CEGB did not say so, but the general opinion was that this first British PWR station would be sited next to the existing CEGB Magnox station at Sizewell, on the Suffolk coast.

The most important nuclear development in Britain in 1979 had, however, nothing to do with the British nuclear establishment at all. On 3 May James Callaghan's Labour government went down to defeat before Margaret Thatcher's Conservatives; and the climate of official opinion was transformed overnight, not least with respect to nuclear power. The effect of this change of climate did not manifest itself immediately in the nuclear context; but the replacement of Tony Benn by David Howell in the post of Energy Secretary presaged a drastic change at least in the flavour of government nuclear power policy. Interviewed by *Nuclear Engineering International* for its October 1979 issue, Howell was asked whether he considered himself 'pro-nuclear'. 'I suppose if you do put it in those terms, I do believe that a future without supersafe, low cost, clean nuclear power is going to be a grubbier, less safe and less decent future.' On first hearing it sounded like a headlong endorsement. On closer examination it was apparent that Howell was careful not to indicate whether he thought that nuclear power already exhibited the attributes listed, or merely that it would have to achieve them. It was an early signal of the approach that the Thatcher Conservatives were going to take to nuclear power policy. Their pronouncements always contrived to sound robustly supportive of nuclear interests; but close textual analysis always revealed built-in escape routes in the event of potential political embarrassment. As it turned out, these escape routes were to carry heavy Ministerial traffic.

David Howell's eagerly-awaited statement to the House of Commons on 18 December 1979 was a masterly example of the escape-route technique. Howell confirmed that the CEGB was to proceed with ordering and constructing one Westinghouse PWR, 'subject to the necessary consents and safety clearances' as always. The following day, the newspapers all reported that Howell had gone on to announce a programme of ten new nuclear stations, totalling 15,000 megawatts, to be ordered at the rate of one a year from 1982 to 1992. He had done nothing of the kind – not, at least, in so many words. What he actually said was this:

Looking ahead, the electricity supply industry have advised that even on cautious assumptions they would need to order at least one new nuclear power station a year in the decade from 1982, or a programme of the order of 15,000 megawatts over ten years. The precise level of future ordering will depend on the development of electricity demand and the performance of the industry, but we consider this a reasonable prospect against which the nuclear and power plant industries can plan. Decisions about the choice of reactor for later orders will be taken in due course.

The nuclear industry seized gratefully on Howell's kind words. Its battered morale needed all the reassurance it could get. But the practical effect of Howell's statement was far from clear. One body that found it unsatisfactory was the newly-constituted Parliamentary Select Committee on Energy. This committee was one of the all-party backbench committees set up to oversee the activities of individual government departments; as such it had no bureaucratic precursor at Westminster. In almost every respect, however, the Select Committee on Energy was the direct lineal successor to the old Select Committee on Science and Technology. It was led by old stalwarts Ian Lloyd and Arthur Palmer, and its membership included several other former members of the earlier Select Committee; and like the earlier committee it opened its innings by weighing into the latest manifestation of British nuclear power policy.

On 30 January 1980 the new Energy Committee questioned its first witness, David Howell; six months to the day later Howell appeared again. In the intervening weeks the committee heard evidence from the CEGB, the AEA, Sir Alan Cottrell, the Nuclear Installations Inspectorate, the NNC, the SSEB, the giant American construction firm Bechtel, Tony Benn, Westinghouse, Arnold (now Lord) Weinstock of GEC, industry trade unions, and a number of independent critics including Friends of the Earth. The dossier of memoranda and transcripts assembled by the Committee made a stack thicker than *Gone with the Wind*; and as the hearings progressed the committee grew steadily more disgruntled. By Howell's second appearance developments – or in some cases the lack of developments – gave the committee an abundance of pointed ammunition.

It had been confidently predicted that early in 1980 the CEGB would issue a letter of intent to the NNC confirming that it wished to order a PWR of Westinghouse design. The letter of intent was at length issued in April 1980. But the CEGB also had to submit to the Secretary of State a formal application for permission to build a PWR. By the end of June no application had yet been submitted. The Conservatives had inherited from their Labour precursors an undertaking that any proposal to build a PWR in Britain would be subject to a public inquiry. One of the preconditions officially acknowledged was that all the necessary safety information and analyses, including those carried out by the Nuclear Installations Inspectorate, would be available to all inquiry participants before the opening of the inquiry. From 1980 onwards this undertaking was reiterated by two Secretaries of State for Energy and by the head of the Health and Safety Executive, parent body of the NII. It fell, in due course, by the wayside. The difficulty even of agreeing on the basic design details of the proposed Sizewell B PWR meant that official safety analysis would still be incomplete when the Sizewell B public inquiry had long since came to an end.

In January 1980 the government announced that the 'three-tier' arrangement of the National Nuclear Corporation, its operating arm, the Nuclear Power Company, and the contract management of GEC was to be rejigged. The three tiers had proved cumbersome and frustrating, with confused lines of communication and responsibility within the companies and between them and their customers, especially the CEGB. For their part the NNC/NPC staff complained bitterly that the CEGB engineers at the Barnwood laboratory in Gloucestershire were duplicating NPC work and making a nuisance of themselves with redundant cross-checking of details.

Although contracts for many components and design features for the Heysham B and Torness AGR stations had long since been issued, the CEGB and SSEB had yet to place the main contracts. The delay was in part due to the CEGB's frank reluctance to surrender its leading role in the construction of a nuclear station costing hundreds of millions of pounds to a company capitalized at only £10 million. CEGB recollections of the débâcle at Dungeness B led it to have understandably strong feelings about the circumstances arising in 1980 and after. What if the NNC made a mess of Heysham

B, or subsequently Sizewell B? Who would pick up the tab? How could the CEGB recover hundreds of millions of pounds in compensation from a company valued at only £10 million? The SSEB too made clear that it intended to retain overall control of its nuclear plant project; and the tug-of-war between the CEGB and its nominal contractor grew steadily more strained and acrimonious. One of the points at issue surfaced during the hearings before the Select Committee on Energy. The Sizewell PWR design was said to be based on the design of the Westinghouse reactor at the Trojan plant in Oregon. However, the Committee was astonished and bemused to learn that the Sizewell design was proving to be more than 30 per cent more expensive than its US equivalent.

On 19 February 1980 the CEGB announced the names of five sites in Cornwall and Dorset that it proposed to investigate as possible locations for nuclear power stations. All the sites were in countryside areas of considerable beauty, and the ensuing public outcry was fierce and hostile. Local people at once gathered into opposition groups; in due course they found themselves in spectacular head-on confrontation with the CEGB, whose public-relations image was growing steadily less persuasive.

The government and the CEGB continued to refer to 1982 as the year in which work on the Sizewell B PWR would commence. But the plausible starting date for actual construction of the PWR was receding month by month. Ron Gausden, Chief Inspector of Nuclear Installations, told the Select Committee in April 1980 that the Inspectorate did not expect to receive a formal application from the CEGB, including design details, for some months; it would then take the NII two years to carry out the required examination before approving the safety of the design. The assertion was an uncanny echo of that made by Gausden's precursor, Eric Williams, before the committee's precursor seven years earlier. In the interim nothing had apparently changed: the NII was still waiting for the requisite information, and the official planners were still ignoring the inevitable impact of the NII's statutory obligations on their confident timetables.

The status of the Heysham B and Torness AGR stations too remained uncertain. Even after Howell's statement of December 1979, confirming the intention to proceed with these orders, both the Cabinet and its Central Policy Review Staff weighed up the possibility of cancelling them. At length, fearing that cancellation might lead to the final collapse of the British nuclear industry, the Cabinet decided to let the AGRs go ahead. Nevertheless, by May 1980 the only contracts placed for either station were design contracts. No hardware was expected to be ordered before August 1980.

A further complication was also entering the picture. Reports in early 1980 revealed that the Nuclear Installations Inspectorate, far from increasing its strength in line with its increasing responsibilities, was actually losing expert staff and cutting back on its work. The NII staff shortage and its implications for the new nuclear programme was even raised in Parliament; on 31 January 1980 Mrs Thatcher responded that she 'was not aware that there was a shortage of staff', and agreed to look into the matter. But no immediate remedial action followed, and the position gradually worsened. NII staff told journalists that their loss of personnel and expertise meant that they could not carry out the necessary safety assessment of the PWR on the schedule put forward by the government. The NII at once formally denied that it was too weak to perform its duties properly; but it conceded that it was 20 per cent understaffed in the London region and was short of expertise in fracture mechanics – precisely the discipline of most concern, relating to the safety of PWR pressure vessels.

The issue received a further nudge on 20 February 1980, when Sir Alan Cottrell appeared before the Select Committee on Energy. Despite all the work that had been done on PWR pressure vessel safety since Cottrell's warnings in 1974 – notably the study by Walter Marshall's team – Cottrell still declared himself 'uneasy' about the safety of the system. He specified three reasons in particular. The high 'power density' of a PWR – its heat output per unit volume – meant that any interruption of cooling would be followed by an immediate and rapid increase in temperature of the reactor core. This gave rise to the first problem. The coolant – ordinary water – had to be kept under a pressure of some 150 atmospheres lest it boil; if for any reason the pressure were to drop – for instance in the event of a leak – the water would flash to steam, and become dramatically less effective as a coolant. The resulting temperature increase might endanger the integrity of the fuel. The second inherent problem was that of the pressure vessel required to withstand the high coolant pressure. Cottrell was still unconvinced that such thick steel could be fabricated and operated with absolute confidence in its integrity throughout the proposed working life of a PWR. The third problem arose from the second: the difficulty of performing adequate repair and maintenance on such a pressure vessel.

Cottrell alluded to the report of Marshall's group, drawing attention to what he called its 'important caveats' especially the 'exacting conditions about standards in workmanship, care in operational control, and rigour in inspection. These conditions call for considerable human abilities. It was beyond the terms of reference of the [Marshall] Study and must be a matter for general judgment to say whether this degree of reliance on human abilities provides an adequately sound basis for the safety of a nuclear reactor.' Cottrell's evidence profoundly impressed the committee; it also profoundly upset Walter Marshall, who had told the committee only a week earlier that he was solidly in favour of Britain building PWRs – that if Britain did not it would be 'the only country in the world driving on the left-hand side of the road and everybody else would have decided to drive on the right-hand side'.

While the PWR programme was being subjected to sceptical scrutiny by the Select Committee, the rival AGRs were also yet again under the unfriendly eyes of the Cabinet. Reports revealed that the government's campaign for cuts in public expenditure might include that for the Heysham B and Torness stations. Despite David Howell's confirmation of the go-ahead for the two stations in his statement less than three months earlier, the Cabinet thereupon asked its Central Policy Review Staff to investigate whether in fact it might not be better to cancel the two AGR stations. Electricity demand was remaining stubbornly static, and the incipient 'programme' to be launched by the new PWR would add yet more redundant generating capacity; surely the AGRs could be dispensed with? However, by mid-March the 'think tank' had returned with the advice – secret, of course, but heavily leaked – that cancellation of Heysham B and Torness might lead to the final collapse of the British power station industry. Even with this unambiguous warning the Cabinet did not at once speak out to dispel the rumours about the AGRs; indeed some reports said that Mrs Thatcher herself – a vigorous proponent of the PWR – was still determined to find some way to stop the AGRs. On

14 April, however, Howell at last told Parliament that the government was giving formal approval for the two AGR station orders. He did, to be sure, note that in the preceding nine months the anticipated cost of the stations had risen from the £2 billion quoted in summer 1979 to about £2.8 billion. Construction was not now likely to begin before August 1980.

On 12 May 1980 the government announced that it had ordered the official Monopolies and Mergers Commission to examine the performance of the CEGB, with particular reference to its finances and efficiency, and to the effect of these on its customers. It was to be a revealing exercise in one respect at least, although the result was not to be made known for a year. Had it been asked, the Monopolies Commission might well also have had something to say about the South of Scotland Electricity Board. On 19 May, in evidence to the Select Committee on Energy, SSEB chairman Roy Berridge conceded that there was now some 60 per cent more generating capacity on the SSEB system than the highest demand it had to meet. Even allowing a generous 25 per cent margin to deal with contingencies like unexpected shut-downs, the SSEB had probably the largest excess of generating capacity in Europe. This did not even include the nearly-completed 1300-megawatt oil-fired station at Peterhead; nor, of course, the Torness AGR station, whose construction had yet to commence.

In its annual report, published in June, the SSEB conceded that the repairs to the Hunterston B2 reactor after the accidental inflow of seawater had cost £15 million; replacement electricity from other plants however cost an additional £42 million. The reactor came back into service in February 1980, after a shut-down lasting more than sixteen months. At the end of June the Nuclear Installations Inspectorate granted the requisite licence for the SSEB to build and operate its new AGR station at Torness. The SSEB said that it expected to commission the first of the two Torness reactors in 1987, and the second a year later. In late June 1980 the estimated cost of Torness had reached £1097 million, compared to an estimate of £742 million in March 1978.

Meanwhile, after a behind-the-scenes boardroom battle, the government announced in April 1980 the appointment of Denis Rooney, a former senior executive of a major private company outside the nuclear field, to succeed Lord Aldington as chairman of the National Nuclear Corporation. Although GEC had asked for and been granted release from its management contract at the NNC its influence was still considerable in the NNC boardroom; and reports indicated that the GEC faction was still at loggerheads with the AGR faction on the NNC board. The battle over his appointment was by no means the last that Rooney would experience during his brief and uneasy tenure as NNC chairman. On 1 July Rooney moved into his office at the NNC, with a memorable comment: 'It's a shambles; it can only get better.' Little did he know. When he was appointed, Rooney was given a brief to carry out the long-debated reorganization of the NNC. When in due course the dust settled, however, it was to be Rooney himself who was carried out.

In late July CEGB chairman Glyn England revealed that the design and safety studies for the Torness and Heysham B AGR stations would not be ready in time for construction to start before the following year, a delay of at least six months. Some industry executives privately anticipated nine. Construction was originally to have begun in February 1980. This had then been put back to August; the further delay was said to arise from a bitter argument about how much the new station designs should depart from those of the first-generation Hinkley Point B and Hunterston B designs. The original intention had been to replicate the earlier stations as closely as possible, to avoid the risks of yet more engineering innovations. Nevertheless, changes accumulated until, according to a report in the *Sunday Times*, there was 'not a single drawing the same'. The CEGB admitted that design changes had already added about 20 per cent to the cost of its new AGR station compared to the cost of Hinkley Point B.

In October 1980 the CEGB at last confirmed what informed opinion had anticipated for many months: the first British PWR would be sited next to the board's existing Magnox station at Sizewell, on the Suffolk coast. It would have a 1200-megawatt Westinghouse PWR. The CEGB would not, however, apply for statutory consent and site licence until 1981; the promised public inquiry would not therefore begin before mid-1982. In the same month, on 7 October, Frank – now Sir Francis – Tombs, a long-time opponent of the PWR, abruptly submitted his resignation from the post of chairman of the Electricity Council. His move came not, in fact, as a consequence of the anticipated advent of the British PWR, but rather because of the abandonment of plans to centralize electricity supply in England and Wales under one mega-authority, with him in charge. Tombs's resignation was to take effect at the end of the year. It injected yet another bubble of uncertainty into a pot already fermenting furiously.

In early December 1980 it was revealed that the Atomic Energy Authority was seconding three of its employees to the under-staffed Nuclear Installations Inspectorate, apparently to write and collate the safety assessment on the PWR. The Institution of Professional Civil Servants, with many members in the nuclear industry and its oversight bodies, deplored the move as 'hazardous and improper'. 'We think that the public needs to realize that, in effect, the industry is attaining a position in which it will assemble the safety assessment on the PWR for itself. That is absolutely contrary to the intention of the Acts under which the Inspectorate was created.' Nor was it a move calculated to reassure the growing band of nuclear sceptics in Britain.

Rumour had been anticipating for some weeks the signing of the actual contracts for the Heysham B and Torness AGR stations. By late December it was however being reported that the signing was unlikely to take place before the end of the year, because of continuing disagreement between the electricity boards and the NNC. The sticking point was still the fundamental question as to whether the NNC was itself the prime contractor, or merely an intermediary between the boards and the plant manufacturers. At issue was the £10 million capitalization of the NNC. This made it unable to indemnify the boards in the event of subsequent difficulties with the two new stations. Furthermore, if the NNC were the prime contractor, a subcontractor who failed to deliver would not inflict damage directly on the NNC, and might escape penalty. Despite the government's expressed desire that the NNC be responsible for total management of nuclear projects, the position seemed untenable – not only for the two AGR stations but also for the Sizewell B PWR. NNC chairman Denis Rooney was reported to be 'disappointed' at this down-grading of the role of the NNC. The CEGB, however, was gradually manoeuvring towards the position it had apparently long desired: that of being itself the main design and contracting body for nuclear plant in Britain.

On 3 February 1981 the CEGB at last submitted to the government and to local authorities its formal application for permission to construct a 1200-megawatt PWR at Sizewell. The formal application triggered the official planning procedure; but the government had yet to make up its mind about the details of the promised public inquiry. It had received a variety of proposals and analyses of planning processes, particularly after the Windscale inquiry of 1977–8, which will be discussed in Part II. However, while the shortcomings and drawbacks of existing procedures had by this time been amply ventilated, no one had come up with any very persuasive alternative. The public, not least the public in the vicinity of Sizewell, had let it be known unambiguously that they would expect a genuine official opportunity to make their opinions heard. But those with unhappy earlier experience of the existing official procedure, especially involving the Windscale issue, were profoundly doubtful about the credibility of such mechanisms as a way to influence government policy.

On 13 February 1981 the Select Committee on Energy published its report on 'The Government's Statement on the New Nuclear Power Programme'. By any criterion the report was a sweeping indictment of current official policy – and yet more so coming from a committee whose members in the main considered themselves strong supporters of nuclear power. The committee began by questioning even the validity of the title of its report. Citing Howell's statement of 18 December 1979, that 'the precise level of future ordering will depend on the projection of future energy demand and the performance of the industry', and that 'decisions about the choice of reactor for later orders will be taken in due course', the committee noted crisply: 'This means, taken at face value, that there is no irrevocable commitment to an ordering programme of 15 gigawatts (and indeed that it is not strictly accurate to describe the announcement as constituting a "programme" at all).'

The committee observed that building 15 gigawatts of nuclear plant 'would represent Britain's biggest public investment programme. We believe it important to stress that this outlay represents a preemption of a large slice of the nation's resources which might otherwise be available for investment in other parts of the economy.' The committee then issued a magisterial rebuke:

> The nuclear 'programme' announced in December 1979 was formulated against the expectation that the winter peak demand on the CEGB system in 1986–87 would be 52 gigawatts, yet within a matter of weeks that forecast was reduced by 7 per cent to 48.5 gigawatts. It would have been less misleading and more helpful to the committee if the CEGB had informed us during their first evidence session that the forecasts contained in their memorandum had already been overtaken by events and were in the process of being revised downwards, even if the precise figures may not have been known at that stage. The credibility of much of the CEGB's subsequent evidence was undermined by this omission and we trust that this will not occur in the future.

The committee reviewed the embarrassing history of the electricity industry's demand forecasting. From serious underestimates in the 1950s the trend had swung definitively to drastic overestimates, starting at the beginning of the 1960s and continuing right through as far as forecasts and actual out-turns could be compared. The July 1974 Electricity Council forecast for peak demand in 1979–80 had been 56.5 gigawatts; actual peak demand had been at most 45.5 gigawatts, including customers that had in fact cut their peak demand at the request of the electricity suppliers. Pointing to the undoubted effects of price increases and economic recession, the committee said that 'further reductions in the CEGB's load forecasts cannot be ruled out'.

Against this background, and weighing possibilities for alternative energy investment, the committee was 'dismayed to find that, seven years after the first major oil price increases, the Department of Energy has no clear idea of whether investing around £1300 million in a single nuclear plant (or a smaller but still important amount in a fossil fuel station) is as cost effective as spending a similar sum to promote energy conservation'. In any case, 'Having examined the economic case for the policy announced by the Secretary of State and, in particular, the figures supplied by the CEGB, we have concluded that many of the underlying assumptions are open to question and that the justification for a steady ordering programme of 15 gigawatts over ten years rests on premises which are necessarily very uncertain.'

After dissecting the CEGB's figures in some detail, the committee found that: 'A number of the assumptions contained in the figures produced by the CEGB and the government in support of their case are, in our view, questionable. Moreover, the methodology employed is in many ways unsatisfactory.' They adduced specific examples of dubious CEGB numerology: the calculation of comparative costs of electricity from different types of plant; the calculated 'net effective cost' of new plant, whose figures 'raise as many questions as they answer about the reliability of the Board's judgment'; the estimation of future cost of fuels; and the estimation of the potential financial impact of delays in construction.

The committee was astonished that the CEGB's figure for the capital cost of a PWR was fully 34 per cent higher than that put forward by the American construction firm of Bechtel for the same plant. The CEGB attributed the difference to the high rate of exchange and 'different industrial factors existing between the two countries'. The committee felt strongly that 'a difference of more than a third seems to us to carry with it unacceptable consequences for the competitiveness of British industry and the general health of the economy'. In conclusion, 'we remain unconvinced that the CEGB and the Government have satisfactorily made out the economic and industrial case for a programme of the size referred to by the Secretary of State in his statement to the House in December 1979'.

It was without doubt a profoundly damaging document – or at least it would have been in any context other than the nuclear. It received extensive coverage in the national press, including the leader columns; and it was seized on by those opposed to the Sizewell B PWR project. But it appeared to have precisely the same effect on official nuclear policy as earlier Select Committee reports: that is, no effect whatever.

On 23 February 1981, on the retirement of Sir John Hill, Walter Marshall took over the chairmanship of the Atomic Energy Authority. In this capacity he also assumed the role of chief nuclear policy adviser to the government. Marshall's accession to this titular leadership was to set in train profound changes, both short-term and long-term, in the organization and management of British nuclear power. It was not, however, to resolve any of the intractable conflicts, either internal or public. Instead, if anything, it was to exacerbate them. Marshall's aggressive and outspoken style of leadership was precisely what the devout nuclear faithful, especially PWR and fast breeder enthusiasts, were longing for; Mrs Thatcher became one of Marshall's most ardent fans. However, this same Marshall approach in short order antagonized many of his own staff and colleagues, and further polarized the existing controversies.

In early April 1981 the NNC delivered its reference design for Sizewell B to the Nuclear Installations Inspectorate. It was now based not on the Trojan plant in Oregon but on the Callaway plant in Missouri. Callaway was one of five identical plants ordered as a group of 'standardized nuclear power plant systems' (SNUPPS). The NNC design was described as having doubled certain key safety systems, and otherwise incorporated major modifications to comply with British nuclear safety standards. Unfortunately, of the five original 'SNUPPS' plants, three had by this time been cancelled, and Callaway was far behind schedule – scarcely an encouraging omen for Sizewell B. Worse was to come. The NII was expected to take about a year to complete its safety assessment of the design; it was understood that the assessment would then be published, allowing three months for its consideration by interested parties before commencement of the public inquiry. It did not, unfortunately, work out that way.

Other nuclear plans were also falling by the wayside. On 19 May 1981, after less than a year on the job, Denis Rooney, chairman of the National Nuclear Corporation, submitted his resignation. He cited 'personal grounds'; but commentators deduced that he was fed up and disillusioned at the failure of the NNC to achieve any genuine standing as an independent and responsible company in its own right. Its meagre £10 million capitalization was still preventing it from assuming credible 'total project management' as sought by the government; and the boardroom friction between GEC and its rivals, Babcock International and Northern Engineering Industries, caught the hapless Rooney in the middle. The corporate partners refused to increase their capital commitments to the NNC, leaving it able to do no more than act as an agent for the generating boards. Rooney's abrupt departure was unexpected and acutely embarrassing for the NNC; but it was all too understandable.

In mid-April CEGB chairman Glyn England had expressed his displeasure that the report on the CEGB prepared by the Monopolies and Mergers Commission had been delivered to government ministers in early March, while England himself had yet to receive a copy. His complaint fell on deaf ears. When the report was at length published, on 20 May, it proved to contain one devastating finding. While broadly accepting that the CEGB's performance in most respects was satisfactory, the commission report included a detailed analysis of recent nuclear investments by the board; and the analysis led to an uncompromising conclusion:

While we find that the Board's demand forecasting has improved, we consider that there are serious weaknesses in its investment appraisal. In particular a large programme of investment in nuclear power stations, which would greatly increase the capital employed for a given level of output, is proposed on the basis of investment appraisals which are seriously defective and liable to mislead. We conclude that the Board's course of conduct in this regard operates against the public interest.

The ensuing row was still echoing a week later, on 28 May, when Norman Lamont, a junior Minister in the Department of Energy, appeared yet again to contradict all the recent disavowals of any commitment to a 'programme' of nuclear stations. Writing in *British Business*, published by the Department of Trade and Industry, Lamont reaffirmed the government's intention to build at least one nuclear power station a year for the next ten years. How he reconciled this intention with the lengthening delay engulfing even Sizewell B, and official insistence that later plants would be ordered only as the need for them became evident, Lamont did not explain.

That some explanation was needed quickly became abundantly clear. Three days later David Fishlock, science editor of the *Financial Times* and a journalist very close to the nuclear establishment, reported that 'A joint review of the reference design for the Sizewell B, Suffolk, nuclear power station will be made by the nuclear industry this summer in what some see as a desperate final effort to prevent the project from falling apart'. According to Fishlock the reference design as delivered – six weeks late – by the NNC threatened 'to offer little or no advantage over the capital cost of the advanced gas-cooled reactor'; in the words of an unnamed CEGB executive, the design was 'too big, too complicated and too costly'. On 14 June the details were duly revealed. Mrs Thatcher herself was reported to have approved the appointment of Walter Marshall to head a 'PWR task force', to take overall control of the project and if necessary bang heads together to get the warring factions to unite behind the scheme.

In short order the new regime made its presence felt. The NNC withdrew the design that it had submitted in April; the design was said to be not only too expensive but also too much of a departure from the 'standardized' Callaway plant in Missouri, on which the NNC design had been intended to be based. The CEGB and the NNC then signed an agreement to exchange information with the four American utilities sponsoring the SNUPPS standardized reactor design scheme. The agreement was to give mutual benefit both to the Americans and the British; no money changed hands. The CEGB and the NNC were then to work jointly on preparation of a new reference design, departing very little from that of the American unit; this new reference design, with accompanying cost estimates, was to be ready for submission to the Nuclear Installations Inspectorate by the end of September. Walter Marshall confidently asserted that he would be disappointed if the new PWR design were not at least 25 per cent cheaper than the new AGRs. How this cost reduction was in practice achieved – what had to be left out to cut the cost – in due course became a focus of urgent concern at the Sizewell inquiry. On 23 July the government published a terse White Paper rejecting the criticisms in the report from the Select Committee, and reaffirming its concept of a 'programme' of nuclear plant – without, to be sure, reconciling the inconsistencies long since obvious. In the same week the government also announced that Sir Frank Layfield, a leading Queen's Counsel at the planning Bar, had been appointed to act as inspector at the inquiry. He did not know what he was letting himself in for.

Meanwhile, although the CEGB and NNC had found a sort of *modus vivendi* for cooperation on the PWR, they were still far apart on the AGR. By September 1981 the CEGB had still not placed an order for the 'nuclear island' of Heysham B. Civil engineering work on the site was long since underway, and much of the other hardware had been ordered; the only part missing was the part that would make it a nuclear station. At last, on 15 October 1981, nearly three years after getting the government go-ahead for the plant, the CEGB signed a £472 million contract with the NNC, for the NNC to provide project management services for Heysham B. It was the first contract for a new power station that the NNC had received since it was set up in 1974, and the first order of a nuclear power station by the CEGB for eleven years. By October 1981 the estimated cost of Heysham B had reached £1.43 billion at March 1981 prices, compared to £1.27 billion a year earlier, at March 1980 prices.

On 20 January 1982 Nigel Lawson, who had become Secretary of State for Energy after Mrs Thatcher had sacked David Howell, made the long-awaited announcement in the House of Commons: the inquiry into the Sizewell B PWR would open a year later, in January 1983. According to Lawson, the CEGB would be publishing its pre-construction safety report on the plant in April 1982, followed by a full statement on the planning application at the end of April. The Nuclear Inspectorate would publish its report on safety issues by the end of June, thus leaving ample time for objectors to study both the CEGB and the NII documents before the commencement of the inquiry. The inquiry, chaired by Sir Frank Layfield, QC, would be held in Suffolk, near the Sizewell site. Cost factors, including environmental costs, would be taken into account; the inquiry would be 'full, fair and proper'. MPs asked Lawson whether funds would be made available to enable objectors to put their cases effectively; Lawson would say only that 'I am quite sure that there will be every opportunity for people to make their case'. Prominent nuclear critics, among them Friends of the Earth, the Council for the Protection of Rural England, and the East Anglian Alliance Against Nuclear Power, soon begged to differ.

Less than a fortnight later yet other critics served notice that the cost question would be high on the Sizewell inquiry agenda. The Committee for the Study of the Economics of Nuclear Electricity (CSENE) was an *ad hoc* panel of independent analysts, chaired by Sir Kelvin Spencer, who had been chief scientist at the old Ministry of Power at the time of the original 1955 White Paper on nuclear power. Spencer had been in his time an enthusiastic supporter of nuclear power; but in the 1970s he realized that 'things had gone very wrong'. In his late seventies he had become a vastly knowledgeable and fiercely outspoken opponent of official British nuclear policy. The committee he chaired had set out to reassess the CEGB's evidence to the Select Committee on Energy and to the Monopolies Commission, and other CEGB assertions about the comparative cost of nuclear electricity and other forms of generation. Its report, released at a press conference in the House of Commons, found that the CEGB's accounting procedures and choice of data were both severely biased in favour of nuclear electricity. The report was scathing. About the Heysham B and Torness AGR stations it declared bluntly that 'Such reactors will, on being commissioned, cost the electricity consumer considerably more than if they had never been built'. The report took issue with the CEGB's use of historic cost accounting in its evaluation of the capital cost of existing nuclear stations, especially the Magnox stations. This allowed the intervening inflation to make the Magnox stations look significantly cheaper with hindsight than they did for the first decade of their lives, and distorted drastically the cost comparison with fossil-fired stations.

The substance of the CSENE report was given poignant immediacy shortly after its publication, by the announcement that the first reactor at Dungeness B – yes, the very same Dungeness B – was about to go critical: to start up for the first time, more than seventeen years after it was ordered. In the passage of time its cost had mounted from the initial estimate of £89 million to over £550 million – well over twice the original cost even after correcting for inflation. The CEGB also conceded that even after Dungeness B began to generate electricity it would be operated far below full power for at least a year. In fact even this announcement of imminent start-up was premature by ten months; but on the timescale of Dungeness B, ten months was a mere bagatelle.

Nor was the future of Dungeness B the only intriguing question in the air concerning the CEGB; the future of CEGB chairman Glyn England was also in doubt. England's defiant defence of the board against the government's financial strictures had made him no friends in the Cabinet. England's contract was to come up for renewal within two months; it appeared ever more likely that the CEGB might soon have its third chairman within five years. When the announcement eventually came it was to be a blockbuster.

On 2 April England called a press conference; the government had indeed told him that his contract, due to end in only five weeks, would not be renewed, and had given singularly feeble reasons for what amounted to his dismissal. His outspoken defence of his industry, and his criticism of Whitehall intervention, left few onlookers in doubt about the real reason for England's abrupt departure, without even a knighthood to his name. Energy Secretary Nigel Lawson was a brusque, abrasive individual who did not take kindly to those disagreeing with his dictates. No replacement for England had yet been found, another indication that the government's handling of the situation was more than somewhat peremptory.

On 20 April the Atomic Energy Authority published a second report on the safety of PWR pressure vessels, once again prepared by the study group led by Walter Marshall. Once again it declared itself satisfied that such vessels would be adequately safe, provided certain conditions were fulfilled, including improved techniques for inspection during construction and in operation, and a means of validating these techniques and those who were to apply them. Given Marshall's new role as task-master of the task force charged with pushing through plans to build the first British PWR, the report might have carried more conviction under a different chairman – for instance Sir Alan Cottrell. Cottrell agreed that the new report reduced his unease about the PWR; he did not, however, go so far as actually to endorse the design.

Reports indicated that the CEGB would publish its statement of case for the Sizewell B inquiry at the end of April. True to form, it did not; but on 12 May, to great fanfare, it did. The CEGB Sizewell project director, Brian George, was photographed sitting smiling between teetering towers of documents, twenty-five volumes weighing in all more than 100 kilograms. The CEGB declared proudly that never before had so much information about a proposed nuclear project been made available to all concerned. John Baker, CEGB member for commercial and public affairs, called the documentation one of the biggest public information exercises of its kind; preparation of the material had cost about £5 million. Only later did it emerge that the massive pile of paperwork nevertheless concealed a gaping lacuna.

In publishing the statement of case the CEGB also said that the estimated cost of the station was now some £1147 million, compared to earlier estimates of the order of £900 million. The additional cost was attributed to extra safety features that had been added to the basic American reference design. The status of these safety features was, however, soon to be called into question, not least in the light of the changes adopted after the April 1981 rejection of the NNC's original Sizewell design.

By this time Glyn England had stepped down as CEGB chairman; but the government had found no one to replace him. His deputy, Fred Bonner, was standing in as acting chairman; but Energy Secretary Nigel Lawson was rumoured to be talking of bringing in someone from outside the electricity supply industry, to give the CEGB what he called an overdue shake-up. On 27 May the government at last unveiled its new CEGB chairman: Walter Marshall. Marshall was reported to be the personal choice of both Lawson and his boss, Mrs Thatcher. They appointed Marshall without consulting either Bonner or his colleagues, a telling augury for subsequent developments in the CEGB boardroom. Marshall took over the CEGB chair on 1 July, with a salary of £51,000 per year, £6000 more than his precursor Glyn England had received. To make the occasion even sweeter for him, he was also granted a knighthood. For those on whom Mrs Thatcher smiled, all was possible. Marshall, to be sure, saw his appointment rather differently. 'I am just an old-fashioned chap who feels it is his duty to serve the country.'

Meanwhile, on 1 June, Sir Frank Layfield, the inspector, met with Sizewell inquiry participants for pre-inquiry discussions on ground rules. The meeting was held in Snape Maltings, the concert hall created by Benjamin Britten for the Aldeburgh Festival, just along the Suffolk coast from Sizewell. The Maltings were to be the venue for the inquiry – an incongruous contrast to the euphony normally to be heard in the converted malthouse. Opponents of the Sizewell application, noting the absurd disparity between their puny resources and those of the CEGB, raised yet again the issue of official financial support for objectors. Lord Silsoe, QC for the CEGB, indicated that the CEGB might itself consider offering financial assistance to objectors. Energy Secretary Lawson had, however, declared robustly that the government would have no truck with such far-fetched notions; and, as most objectors glumly anticipated, representations by Layfield were to be unavailing.

Few, however, anticipated one other development at the pre-inquiry meeting. Despite the CEGB's triumphant publication of its statement of case a fortnight earlier, it emerged that the 100 kilograms of paper did not in fact include the final version of the pre-construction safety report for the Sizewell B PWR. The Nuclear Installations Inspectorate was still reviewing the draft safety report given to it – late, as usual – by the CEGB in December 1981. The final report would not be ready until the NII had given its blessing.

An NII representative at the June meeting at Snape gave an assurance that the NII review would be published by 15 July. When the review was duly published, far from endorsing the CEGB's safety report it revealed that the NII was still unhappy about five major areas of concern. Objectors pointed out that the need for further design work meant in turn that they would not have the final documents in time to give them adequate examination before the inquiry opened. This proved to be a drastic understatement. Four months later, on 30 November, Chief Nuclear Inspector Ron Anthony, introducing the NII annual report, said that the NII was still unsatisfied about several safety questions concerning Sizewell B, and had asked the CEGB to provide further evidence. The NII would not issue the necessary site licence until it was fully satisfied about all aspects of the design. What this would do to the inquiry schedule, and the repeated official undertakings about a 'full, fair and thorough' inquiry, was all too easy to deduce.

On 20 November 1982 the CEGB announced that it was awarding the French firm of Framatome the design contract for the pressure vessel for the Sizewell PWR; the manufacturing contract would follow if the inquiry gave the project the go-ahead. Back in Britain, however, the interminable boardroom warfare between rival factions was breaking out yet again. Both GEC and Northern Engineering Industries, the company formed by the merger of Clark Chapman and Reyrolle Parsons, were represented on the board of the National Nuclear Corporation. But the two companies were each determined to win the design contract for the Sizewell B generating sets. The design contract itself was worth only some £2 million; but both companies believed that it would be followed by a manufacturing contract potentially worth £100 million, to say nothing of later repeat orders for future PWRs. Accordingly the two factions were up to their old tricks – the ones that had cost Denis Rooney his chairmanship six months earlier. The CEGB for its part had rejected both initial tenders and called for resubmission. The final decision, expected in January 1983, would be taken at Cabinet level; the lobbying grew ever more strenuous. As usual, a nuclear power decision was to be based far more on political clout than commercial rationality.

On 24 December 1982 the CEGB received a Christmas present of sorts, albeit some twelve years late: the first reactor at Dungeness went critical, for the first time. It did not, however, produce power. After such a long gestation the site team intended to nurse it gingerly to life; it would not be hot enough to raise steam for the generating set for another fortnight. Unfortunately, when it did, it at once suffered a breakdown putting it back out of operation for yet further protracted surgery.

On 11 January 1983 the Sizewell B inquiry opened for business in the Maltings at Snape. It was already a year after the date the government had foreseen in December 1979 for actual commencement of construction of the plant. The inquiry was expected to last up to six months. By the time it finished in the spring of 1985, more than two years later, it had done for public inquiries what Dungeness B had done for nuclear power stations.

PART II
Reprocessing an Obsession

5 The ambiguous atom

For the first decade of British nuclear activities uranium was not a fuel, and reactors were not for generating electricity. That was all to come later. In the 1940s and early 1950s uranium was bomb-material, and reactors were a way to turn uranium into a better bomb-material. At the time uranium was rare and costly; it was also, for obvious reasons, acutely sensitive politically. Not only was it hard to come by: even when you had it, more than 99 per cent of it was useless. Only seven uranium atoms out of every thousand – 0.7 per cent – were the lighter kind called uranium-235, that would sustain a nuclear 'chain reaction', or explode. The remainder were uranium-238; and the two kinds ('isotopes') of uranium atom were chemically identical. Sorting out the potentially explosive uranium-235 from its vastly more abundant but unusable heavier sibling was extraordinarily difficult.

There was, however, another way to produce nuclear bomb-material from uranium. If ordinary uranium could be made to undergo a controlled chain reaction, in a 'nuclear reactor', the reaction would convert some of the useless uranium-238 into another substance, called plutonium. Plutonium-239 was an even better bomb-material than uranium-235. Furthermore, because plutonium was chemically a different substance from uranium, plutonium could be separated chemically from uranium with comparative ease – certainly more readily than uranium-235 could be separated from uranium-238.

Accordingly, the handful of people who created Britain's nuclear weapons programme from 1946 onwards decided to build reactors that could convert a fraction of ordinary uranium into plutonium. Between 1948 and 1952, a team of scientists, engineers and site workers led by Christopher (later Lord) Hinton designed and erected two massive plutonium-production reactors at a site on the north-west coast of Cumberland, which they called – for reasons now apparently forgotten – Windscale. The reactors contained an enormous quantity of uranium – secret, but probably hundreds of tonnes; however, it was not 'fuel' as the term is normally used. Indeed the heat given off by the nuclear 'chain reaction' in the uranium was not a desired output, but a severe engineering problem.

The plutonium-producers had, moreover, one further major problem. The uranium metal was inserted into a reactor, and the reactor was started up. After some weeks, the reactor would be shut down and the 'irradiated' uranium removed. Having undergone the chain reaction the metal rod was no longer pure uranium. Some of its atoms had been 'split' into lighter fragments called 'fission products', for instance strontium-90 and caesium-137. Some of the useless uranium-238 had been changed into plutonium-239. Depending on how long the chain reaction had run, some of this plutonium-239 might also have been further changed into plutonium-240, 241 and 242. The fission products were variously radioactive: some of them intensely so, and some for a considerable time – decades or even centuries. To recover the plutonium for use in a bomb involved chemical separation of these different constituents. The chemistry itself is not very complex; but it was complicated by the radioactivity, especially that from the fission products. The separation procedure was called 'reprocessing'.

While the scientists and engineers were constructing the Windscale reactors, they were also designing and constructing the chemical reprocessing plant that would be required for recovery of the plutonium produced in the reactors. The reprocessing plant was built not far from the reactors; it was designated Building 204, or B204 for short. B204 produced the plutonium used for Britain's first nuclear weapons test, on 3 October 1952, off the Monte Bello Islands near the northwest coast of Australia.

When the Calder Hall reactors came into service in 1956, they were, as mentioned in Part I, dual-purpose reactors. Their primary purpose was to augment the supply of weapons-plutonium; but they were designed with circuits that allowed the heat of the chain reaction to be collected and used to raise steam and generate electricity. This by-product electricity was the basis for calling Calder Hall 'the world's first nuclear power station'. From this viewpoint it was correct to call the irradiated uranium from Calder Hall 'spent fuel'; its heat output had been used. But this 'spent fuel' was actually the primary product from Calder Hall; its 'fuel' value was 'spent' but its military value was yet to be realized.

The spent fuel was discharged from Calder Hall and transported in massive shielded 'flasks' to be stored in water-filled 'cooling ponds' on the adjoining Windscale site. When the uranium had cooled enough – a few weeks or months – it was fed into B204 and reprocessed: separated into unused uranium, plutonium and fission products. The fission products, in the form of hot, fiercely radioactive solution, were piped to another building nearby to be stored in shielded stainless-steel tanks. The plutonium and uranium were purified and sent for further use.

It has, needless to say, always been difficult if not impossible to learn very much about the operation of the B204 reprocessing plant. Its crucial military significance for British nuclear weapons made it even more hush-hush than most British nuclear activities. It appears, however, to have functioned as intended, and served its military purpose well. No official figures have ever revealed its cost; but its influence on subsequent nuclear thinking, in the very different civil context, must not be underestimated. Because of its role in reprocessing the spent fuel from the dual-purpose Calder Hall reactors, and their northern cousins at the Chapelcross station, similarly dual-purpose, B204 helped to establish the direction of British nuclear thinking about what to do with spent fuel from any reactor of whatever kind or purpose. What did you do with spent fuel? You reprocessed it. Through the years this assumption continued to prevail, even though the context and the circumstances gradually changed almost beyond recognition.

By 1958 the Atomic Energy Authority was laying plans for reprocessing of the spent fuel not only from the new Calder Hall and Chapelcross stations – eight reactors in all – but also from the Magnox stations of the first 'commercial' nuclear programme, at Berkeley, Bradwell, Hunterston and other sites yet to be decided. These stations would produce much more spent fuel than the B204 plant could handle; and the fuel would be significantly different in one important characteristic. Instead of being removed from the reactors after at most a few months, it was designed to remain in them for more than a year, receiving a substantially higher 'burn-up'.

It would therefore contain substantially more fission products, making it substantially more radioactive. The AEA designers set to work to create a plant to reprocess this commercial Magnox fuel.

They took for granted that this fuel, like that from the weapons-reactors, would have to be reprocessed. The assumption even governed the design of the commercial Magnox stations themselves. Each station was built with a water-filled 'cooling pond', into which spent fuel would be put after its removal from a reactor. The Magnox alloy in which the uranium metal fuel rods were canned served adequately to protect the uranium from the carbon dioxide coolant gas in the reactor. But Magnox alloy corroded rapidly in water. The cooling ponds were intended to be only an interim stopping-place for the fuel, to allow its temperature and its radioactivity to fall. After perhaps 150 days in a station pond it was then to be shipped to Windscale for reprocessing. The new reprocessing plant at Windscale included a reception point for shielded steel flasks of spent fuel, a storage pond for the open baskets of fuel rods, a building for stripping the Magnox cladding off the fuel rods, designated building B30, and the new large chemical separation plant, building B205.

According to the laconic information provided by the sixth annual report from the Atomic Energy Authority, 1959–60, B205 was 'due to be in operation in 1963'. It wasn't, at least not in 'full active operation', in the words of the tenth annual report; it achieved this latter status in mid-1964. The specialized vocabulary of nuclear engineering includes an extensive lexicon of terms to describe the status of plants that are to be sure not entirely inert, but on the other hand are not actually doing what they have been built to do. The blanket word 'commissioning' serves to cover a multitude, not precisely of sins but of embarrassments – for instance the five-year 'commissioning' of the Wylfa nuclear station. To the uninitiated the term 'in operation', as used in the earlier AEA report cited above, means 'doing its job'. In a nuclear context this is not necessarily so.

The B205 reprocessing plant was commissioned in June 1964, to the evident satisfaction of the AEA: 'Commissioning of the plant was accomplished remarkably quickly, and during even the first campaign [batch of reprocessing] rates were achieved which were well in excess of the design capacity.' It sounded impressive, until the operators of Windscale revealed more than ten years later just how elastic this 'design capacity' actually was. Meanwhile, even before the radioactivity in the old B204 plant had cooled down, the AEA began making new plans for it. What became of these plans will be described in Chapter 6.

The B205 Second Chemical Separation Plant had been constructed to replace the B204 First Chemical Separation Plant. One of B205's jobs – the one that was mentioned officially – was to increase reprocessing capacity at Windscale to cope with the new commercial Magnox stations. The other, less publicized job was to do what B204 had been built to do: to recover plutonium for nuclear weapons. The B205 reprocessing plant, like the Calder Hall and Chapelcross reactors, was a military facility masquerading as a 'commercial' one. No details about the financial arrangements covering the weapons-related activity of the B205 plant have ever been published. It has now been in service for more than twenty years; and its status is as ambiguous in 1985 as it was in 1964. No other aspect of reprocessing in Britain is as dubious, or as diplomatically explosive.

The CEGB, which until 1965 had remained sturdily sceptical about nuclear power, thereafter became, as described in Part I, a headlong proponent. In so doing it accepted the received wisdom – received in the main from the AEA – that spent fuel had to be reprocessed. As it began discharging spent fuel from its Magnox reactors into cooling ponds at its power stations, it contracted with the AEA to ship the spent fuel to Windscale, there to be reprocessed in B205. Since the AEA was now providing a variety of such services – fuel manufacture, uranium enrichment and reprocessing – not only to the British electricity boards but also to foreign customers, the government in 1965 reorganized the AEA. Fuel services became part of a new 'Trading Fund', no longer financed unhesitatingly by direct vote of Parliament. Instead, these services were to be operated on a commercial basis, and keep separate accounts, with services to be purchased at a price that would cover their costs and indeed enable the AEA to build up financial reserves for future investment. Allowing for the ambiguity of the financial provision for reprocessing – for weapons as well as putative civil objectives – the concept was clearly reasonable. In practice it led to a series of progressively more stubborn confrontations between the CEGB and its monopoly fuel-service supplier, especially about reprocessing.

Minor teething troubles aside, the B205 plant operated throughout the rest of the 1960s as intended. Modifications even increased its capacity. The discharges of spent Magnox fuel from the commercial stations of the CEGB and the SSEB grew steadily more copious; but B205 seemed to be able to cope. All was not, however, quite as well as it seemed. The AEA's fuel-manufacturing plant at Springfields had come up with modern designs of Magnox fuel elements, more durable and with better heat-transfer characteristics. These new elements could be left in a reactor significantly longer, to produce significantly more heat-output before they had to be discharged. The advantage of this higher 'burn-up' brought with it, however, a subtle disadvantage that went unnoticed for several years. When at last it did come to light it was to put a major crimp in the hitherto trouble-free record of B205.

In the meantime, Britain's interest in reprocessing also acquired several international dimensions. Spent fuel from the British-supplied Magnox reactors at Latina, in Italy, and Tokai Mura, in Japan, was shipped to Windscale for reprocessing. The separated plutonium from this fuel was thereafter returned as a matter of course to the customers in Italy and Japan – a practice that might have attracted more comment than it did. Plutonium was potential nuclear weapons material. Britain had nuclear weapons; but neither Italy nor Japan did. In 1968 Britain joined with the US and the Soviet Union in co-sponsoring the landmark Treaty on the Non-Proliferation of Nuclear Weapons. The Treaty came into force on 5 May 1970. Article III of the Treaty committed member countries not to supply nuclear technology or materials to any non-nuclear-weapons country that did not accept full-scope Treaty safeguards on all nuclear activities in the country. Britain, as one of the Treaty's three depositary countries, might have been expected to observe this undertaking with special scrupulousness. But Britain continued to supply separated plutonium to both Italy and Japan, for five years and six years respectively, before either customer country actually ratified the Treaty or accepted its safeguards.

The AEA also had a small reprocessing plant at its Dounreay site on the north Scottish coast, primarily for reprocessing spent highly-enriched uranium fuel from the Dounreay Fast Reactor and the nearby Materials Testing Reactor. But the AEA also carried out contract reprocessing of spent fuel from Federal Germany, Denmark, Japan and Canada in the Dounreay facility. The radioactive waste material from this foreign spent fuel was stored with that from British spent fuel, on the Dounreay site.

By the end of the 1960s the AEA was setting its international reprocessing sights yet higher. Its new Head End Plant, the converted B204 – to be described in the next chapter – spurred its imagination. By this time the AEA's fuel-service activities constituted a substantial business in their own right. In recognition of this the Labour government under Harold Wilson prepared legislation to separate the fuel business from the AEA's more research-oriented aspect. The Labour government, to its surprise, was swept away by the general election of June 1970, and the AEA legislation with it. But Edward Heath's Conservatives at once resurrected the legislation, and in 1971 it was duly passed. It broke the AEA into three separate entities; the two new ones were The Radiochemical Centre, in Amersham, Bucks, and British Nuclear Fuels Ltd, with headquarters at Risley in Lancashire. The nominal independence of the two new organizations was to be sure slightly overshadowed – as mentioned in Part I – by the fact that all of their shares were owned by the AEA, and that the chairman of each of the new organizations was Sir John Hill, chairman of the AEA. British Nuclear Fuels Ltd (BNFL) took over all the fuel-service installations of the old AEA: the uranium and fuel-manufacturing plant at Springfields in Lancashire, the enrichment plant at Capenhurst in Cheshire, the dual-purpose Calder Hall and Chapelcross uranium-plus-electricity stations, and the entire Windscale complex with the exception of the small Windscale prototype AGR.

Despite its new status as a 'commercial' company BNFL continued to carry out one not strictly 'commercial' activity. BNFL took over the AEA's responsibility for the production of fissile material for British nuclear weapons, using the reactors at Calder Hall and Chapelcross and the B205 reprocessing plant at Windscale. The financial basis of this activity was kept out of the 'commercial' books, as indeed was virtually any public acknowledgement of it. In due course this 'civilitary' status of Windscale also profoundly complicated – not to say discredited completely – Britain's attempt to establish itself as a responsible party to the international system of nuclear 'safeguards' intended to keep civil nuclear activities on the civil straight and narrow.

By the time that BNFL took over Windscale, on 1 April 1971, the commercial Magnox stations were discharging spent fuel with a burn-up of over 4000 megawatt-days per tonne of uranium, compared to 3000 in the early years of the Magnox programme. The B205 reprocessing plant could cope quite adequately with this more radioactive spent fuel; and for a time all seemed well. The plutonium and unused uranium were recovered as usual; the intensely radioactive fission products, in solution in hot nitric acid – so-called 'high-level liquid waste' – were piped from B205 to building B215 for storage, in double-walled stainless steel tanks with complex cooling systems. By the summer of 1972, however, something untoward had begun to happen. In order to reduce the volume of high-level liquid waste per tonne of spent fuel reprocessed, BNFL staff allowed the liquid waste to boil for a time under the heat of its own radioactivity, to lower its water content. Unfortunately the latest batches of spent fuel presented a problem. During their longer sojourn in power station reactors, the fuel elements were subjected to a significantly larger total amount of neutron radiation. As a result some of the Magnox magnesium alloy from the cans enclosing the fuel rods migrated into the outer surface of the uranium rods themselves. When the cans were mechanically stripped off the rods at Windscale, this thin layer of magnesium alloy remained on the surface of the rods, and was dissolved in the nitric acid with the rest of the rods.

The compounds it formed in the high-level liquid waste included some that were substantially less soluble than those from the spent fuel itself. In consequence, BNFL staff were unable to let so much of the liquid boil itself away: if the solution were too concentrated some of the Magnox compounds would come out of solution and settle to the bottom, depositing a layer of fiercely hot solid radioactive material on the floor of the tank, and endangering its long-term integrity. The unexpectedly large volume of waste thus requiring storage as liquid soon overtook the programme of construction of new storage tanks. It was company policy always to have one extra empty tank available in case a tank in use had to be emptied; but by late summer 1972 there was only one tank not in use. Accordingly, in September 1972 BNFL had to suspend reprocessing and shut down B205.

The plant remained shut down until the following summer, by which time further tank storage capacity had been completed. In the meantime, however, spent Magnox fuel had been accumulating in the storage pond at Windscale. As mentioned earlier, the outer cladding of Magnox fuel was designed to protect the uranium rod against carbon dioxide coolant in a reactor. It was not designed to survive in the much more corrosive environment of a water-filled cooling pond; after immersion of a year or so the Magnox might begin to deform and leak. By the time B205 came back into service in 1973 some of the older fuel in the Windscale pond was deteriorating significantly. Its degenerating physical condition not only raised the level of radioactivity in the pond area; it also made the fuel more difficult to reprocess. Although work began at once to clear the backlog, fresh spent fuel kept arriving from the Magnox stations around the country, and reprocessing in B205 fell farther and farther behind. Ere long the ponds at CEGB and SSEB stations were also beginning to fill with ageing spent fuel; and this fuel too was beginning to deteriorate.

BNFL said nothing whatever about these problems in public. Nevertheless, by the mid-1970s the situation was almost out of hand. Discharges of radioactivity from the Windscale pond into the Irish Sea had increased dramatically. BNFL had unilaterally begun refusing to accept any further shipments of spent fuel from the commercial Magnox stations, despite its contractual agreements with the electricity boards. In January 1975 BNFL shut down the first of its reactors at Calder Hall. Calder Hall itself had no cooling ponds; fuel discharged from the reactor was normally taken across the little Calder river and stored in the Windscale pond. But this time the Calder Hall reactor was simply left shut down but pressurized, with the spent fuel left inside it. The electricity sent out from Calder Hall to the national grid was one of BNFL's main money-earners; the industry press was obviously curious to know why one of the reactors was shut down. But BNFL refused to explain, even to the much-respected newsletter *Nucleonics Week*. The company insisted that the shut-down was merely a temporary measure of no particular significance. However, the first Calder Hall reactor thereafter remained shut down for more than five years. In February 1976 reports revealed that the Bradwell station in Essex had 9000 spent fuel elements in a pond whose maximum capacity was supposed to be 2000; and the station staff were irate.

Similar problems were also arising at other commercial Magnox stations around the country. Not until mid-1977 was the whole embarrassing story of the Magnox reprocessing shambles pieced together, in a painstaking cross-examination of BNFL by Friends of the Earth, at a major public hearing called the Windscale inquiry, to be described in Chapter 7.

By that time, however, BNFL had already applied for, and been granted, government investment approval for what the company called 'refurbishment' of the Magnox reprocessing facilities at Windscale. The application involved obtaining government sanction for an increase in the amount BNFL could borrow – with full government backing for the loans, needless to say. The application came before the House of Commons, and was heard by a committee of back-benchers; it received the go-ahead, as the Nuclear Industry (Finance) Act, in the summer of 1976. If MPs had known about the confusion then reigning at Windscale they might have pressed BNFL's witnesses to be a little more forthcoming about the planned 'refurbishment', then estimated to cost £245 million. Not until the following year did BNFL reveal that this 'refurbishment' entailed the construction of an entire new Magnox plant, larger than any other building at Windscale, on a completely new part of the site. Even by the normal standards of nuclear industry syntax it was a remarkably elastic interpretation of the word 'refurbishment'.

Sorting out the Magnox mess would have been a major task on its own terms. The backlog of spent fuel was not cleared until the end of the 1970s; and even by 1985 the new Magnox facilities at the Windscale Works were still not completed. Nevertheless, throughout this entire period BNFL kept its Magnox reprocessing difficulties under cover. Instead, it pressed on eagerly to embroil itself in an even more demanding task. The advanced gas-cooled reactors and light-water reactors in modern nuclear power stations used fuel made not of uranium metal but of ceramic uranium oxide. This oxide fuel was much more durable than metal fuel; it could stay in a reactor for several years. It therefore produced much more output per fuel element; but it also therefore at length emerged from the reactor some ten times as radioactive as metal fuel. BNFL's first venture into reprocessing oxide fuel was to be anything but encouraging.

6 Heading for a fall

In the mid-1960s, soon after the B205 reprocessing plant came into operation, it became evident that the plant would have more capacity than the existing Magnox programme would require. It could not, however, cope directly with the spent uranium oxide fuel that would be discharged from the advanced gas-cooled reactors, nor the similar fuel already beginning to emerge from light-water reactors elsewhere. Right next to the new B205 plant was the old B204 plant, now retired from service, still heavily contaminated but structurally sound. BNFL staff drew an obvious conclusion: they proposed to decontaminate B204 and convert it into a plant for pre-treating spent oxide fuel, by chopping it up and dissolving it ready to feed into the adjoining chemical separation plant in B205.

The decontamination and conversion took several years; it also entailed demolishing some 300 cubic metres of reinforced concrete shielding, using in all about a tonne of high explosive inside the building. New hardware was developed, installed and tested under non-radioactive conditions. At last, in August 1969, the first real spent oxide fuel, from the nearby Windscale prototype AGR, was fed into B204, and thence in solution via a collection of buffer-tanks into B205. The AGR fuel was followed by a batch from the Garigliano boiling-water reactor in Italy; and the converted plant functioned exactly as intended. BNFL christened it the Head End Plant. Its nominal annual capacity was said to be 300 tonnes of spent oxide fuel. This was, it is true, contingent on the availability of B205 to carry out the actual chemical separation after the fuel had been dissolved in the Head End Plant. But the Magnox programme had already run to its conclusion; no more Magnox reactors would be built. Furthermore, the spent fuel from the Magnox stations was already achieving higher burn-up; less fuel would be discharged for the same station output, leaving a larger fraction of B205's operating-time available for use with the Head End Plant.

The implication was clear. Foratom, an international nuclear industry trade organization, spelled it out in a report called *The Future of Reprocessing in Europe*, published in the spring of 1970. According to current plans, not only in Britain but also in France, Federal Germany, Belgium and elsewhere, reprocessing capacity was likely to expand far beyond the plausible requirements of the nuclear power programmes in Europe. The report noted that talks were already underway between the reprocessors in Britain, France and Federal Germany, seeking a tripartite agreement about coordinating their expansion plans. Among the plans cited was that of the British to double the capacity of the Windscale Head End Plant by 1975 – presumably still in conjunction with B205 to carry out the actual chemical separation of the dissolved oxide fuel. The Foratom report warned that over-expansion would lead to economic difficulties for the reprocessors. It noted that the large existing plants in Britain and France benefited from lower construction costs at the time they were built, and from the 'baseload' of uranium metal fuel already being reprocessed in the plants. New plants being planned would have neither of these advantages; nor could they rely on government subsidies like the small experimental plants in Belgium and Federal Germany. The existing surplus capacity was already complicating commercialization of reprocessing: 'It is not surprising that the utilities attempt to take advantage of this competitive situation by signing long-term contracts. This provides a risk that prices will remain at this unhealthy low level and will depress capital investment, since new large plants will require a considerable increase.'

Within a year the British, French and German reprocessors had taken the message to heart. In mid-1971 they formed a joint company, United Reprocessors GmbH, registered in Federal Germany, 'for the purpose of marketing and providing services for reprocessing of irradiated fuel from nuclear power stations using uranium oxide fuel, including the transport of irradiated fuel and recovered products and the conversion of recovered products'. In some respects this new company looked uncomfortably like a cartel, whose main purpose was to restrain otherwise damaging competition between reprocessors in the three partner countries. As it happened, however, the point was to prove largely academic.

According to the industry monthly *Nuclear Engineering International* for November 1971, 'The master scheme of United Reprocessors is that the Windscale plant in the UK' – that is, the Head End Plant – 'will meet the main demand for oxide reprocessing through to 1977–78 with the existing plant capacity of 300 tonnes per year stretched to 400 tonnes per year and then doubled up to 800 tonnes per year in 1976 . . .' with French and German plants then joining in by 1981. The magazine went on to say:

> The trouble with this master plan is that it gaily assumes that United Reprocessors are going to get a virtual monopoly of the European business. There may be some justification for this belief in the case of the UK and French markets where all the parties concerned are government controlled but the private utilities in Germany may have very different ideas especially if they are offered attractive prices from the much larger plants which will have a huge surplus of capacity in the USA after 1975.

As prognostications go this one got almost everything wrong but 'a' and 'the'. Far from there being a 'huge surplus of capacity in the USA after 1975' there was no civil reprocessing plant operating at all; indeed in 1985 there still isn't. In short order the private utilities in Federal Germany had to set up their own company to do their reprocessing; and in 1985, fourteen years later, it has yet even to start building a plant. As for the UK, what went wrong was comprehensive; but it had nothing to do with anyone else taking away the business. The vast expansion of reprocessing capacity anticipated by the 1970 Foratom report failed utterly to materialize. In Britain, far from building on the putative success of the B204 Head End Plant, its operators were soon to disown it entirely.

In mid–1971, as described earlier, the operators in question became British Nuclear Fuels Ltd. Nevertheless, apart from the change of corporate name, everything at Windscale remained much the same, including the expansion plans. But the plans were running into some modest difficulties. The spent oxide fuel from British advanced gas-cooled reactors had not yet begun to arrive; indeed it now seemed unlikely to do so for several years. On the other hand, contracts for reprocessing spent oxide fuel from foreign customers were being signed, and consignments of foreign fuel were already stored in the Windscale ponds, with more on the way. The Head End Plant performed as intended, when it could be used. Unfortunately, by mid-1972 the aforementioned trouble with storage-tank construction put the B205 plant out of action for nearly a year. The Head End Plant was therefore also unable to function. Despite its nominal capacity of 300 tonnes of spent fuel per year, in its four years of operation to September 1973 the Head End Plant had reprocessed only 100 tonnes of fuel in all.

By the time B205 started up again in the summer of 1973 it had to be earmarked primarily for work on the backlog of deteriorating Magnox fuel. Nevertheless, on 26 September 1973 BNFL staff were engaged in preparing the Head End Plant for a fresh 'campaign' of oxide reprocessing. Unknown to the staff, a long-gestating hidden problem was about to manifest itself all too dramatically. The higher burn-up of uranium metal Magnox fuel had already produced unexpected and embarrassing side-effects, of an insidious kind, as described earlier. The higher burn-up of uranium oxide fuel also had a surprise in store, this time abrupt and unnerving.

During the chain reaction in the core of a reactor, fission product materials accumulated in the fuel. Some of these fission products were isotopes of rare metals like rhodium – intensely radioactive and essentially insoluble, even in hot nitric acid. If the fuel remained long enough in the reactor, these insoluble fission products coalesced into tiny granules. When the spent fuel was eventually dissolved, the granules were not. Instead they were carried along in the liquid process stream until they reached a place where they could settle out. During previous operation of the Head End Plant, a layer of these granules had formed on the bottom of a process vessel. Their searing radioactivity had long since evaporated all traces of liquid from the floor of the vessel, and left it scorching hot. As the first flow of the new batch of process fluid poured on to the overheated floor of the vessel, it produced a steam explosion – a sudden violent evaporation accompanied by a sharp pressure-surge that carried a puff of radioactivity out through the shaft seal on the vessel, past the shielding and into the air of the room where the staff were working.

Alarms sounded immediately. However, as the subsequent painstaking investigation by the Nuclear Installations Inspectorate revealed, the alarms went off frequently for no reason at all; and most of the staff ignored them. The original B204 separation plant had been designed to operate mainly by gravitational flow, with as few moving parts like pumps as possible. The building was ten storeys high – but it had no personnel lifts. When senior staff realized that the alarms were genuine, and that there was radioactivity in the working environment of the plant, they had to run up and down the ten long flights of stairs, shouting to their colleagues to clear the building. Doing so took about half an hour; the last two employees were found and warned only after a frantic search.

Following prescribed emergency procedures the staff assembled at the medical centre, where they were examined for radioactive contamination. In all thirty-five workers were contaminated, mostly with the isotope ruthenium 106. The Nuclear Installations Inspectorate at once embarked on a meticulous study, culminating in a report revealing the sequence of events described above. The NII report did not, however, appear until a year later. In the intervening period BNFL were almost blithely unconcerned about the 'incident' – so much so that it did not even qualify for mention in the company's next annual report, for 1973–4. BNFL chairman Sir John Hill did not so much as suggest that the Head End Plant had been put out of service, notwithstanding that by the time of the 1973–4 annual report in September 1974 the plant had been shut down virtually a whole year. The following annual report, for 1974–5, was not much more forthcoming:

> Reprocessing of the more highly irradiated oxide fuels as used in light water reactors is now generally recognized to be an exacting task. Operations on the 'head end' plant at Windscale for handling such fuels have had to be suspended until such time as plant modifications have been completed, but the experience gained is being applied to the new plants now being planned.

Coming two years after the incident in the Head End Plant, and with the plant still shut down, the chairman's comment exhibited striking insouciance, to say the least. As it turned out, the 'suspension' was permanent, and the 'modifications' terminal.

To be sure, BNFL gave no indication of this embarrassing possibility for nearly four years. But among themselves they had apparently, well before 1977, laid the unfortunate Head End Plant to rest. They wished only to forget it, all ten storeys of it. They had much more ambitious plans.

7 Helping with inquiries

Even while the radioactive dust was still settling in the B204 Head End Plant, BNFL planners were looking beyond its shadow to what they were sure would be the sunlit uplands of oxide reprocessing. A year after the B204 incident, while chairman Sir John Hill was ignoring it in his annual report, BNFL chief executive Dr Norman Franklin was telling his staff that the company was making plans for a new and much larger oxide reprocessing plant – indeed possibly even two. Their capacity might be 1000 tonnes of spent fuel per year each. One plant would service British domestic power stations that used oxide fuel – at the time the AGR stations. This domestic plant, to be in operation early in the 1980s, would cost between £100 and £200 million. The other plant would be dedicated to serving foreign customers. By this time, of course, BNFL already had a substantial inventory of foreign spent oxide fuel stored in the Windscale ponds. It had been delivered for reprocessing via the Head End Plant, under contracts signed before the Head End Plant ran aground. These contracts indeed had some time to run, and more shipments of fuel from outside Britain were still arriving, although BNFL by this time had no plant in which to reprocess it. United Reprocessors remained sanguine; according to them the next 'campaign' of oxide reprocessing in the Head End Plant would start in late 1975, after completion of the modifications recommended in reports on the September 1973 incident. Meanwhile, in November 1974, BNFL began preliminary discussions with the Enrichment and Reprocessing Group of the nine Japanese electricity suppliers, discussing possible terms for reprocessing Japanese spent fuel in the 1980s.

These activities went, as usual, totally unnoticed by politicians and the public, until mid-January 1975. Then, within one twenty-four-hour period, two Windscale process workers died, one of leukaemia and one of myeloma – diseases known to be caused by radiation. The media picked up the story; and the ensuing uproar forced BNFL to arrange the first press-visit to Windscale since the inauguration of the Windscale AGR in 1962, thirteen years before. At the press conference on the site on 30 January 1975 most media interest was focused on questions of radiological safety. One correspondent, however, asked Dr Franklin if it was true that BNFL was still supplying separated plutonium to Italy and Japan, although neither country was then a party to the Non-Proliferation Treaty, of which Britain was one of the three depositary countries. Franklin responded that this question would have to be addressed to the British government. The direct answer would have been 'yes'; such nuclear export activities – even those in contravention of Treaty obligations – were at the time undertaken with complete freedom from public or Parliamentary oversight. This comfortable arrangement was, however, on the verge of transformation.

In May 1975 the environmental organization Friends of the Earth published a four-page tabloid called *Nuclear Times*. It described BNFL's plans to expand reprocessing at Windscale, which would make the site 'one of the world's main radioactive dustbins'. At the time the story attracted little attention. On 21 October 1975, however, the *Daily Mirror* splashed its front page with a stark headline proclaiming 'PLAN TO MAKE BRITAIN WORLD'S NUCLEAR DUSTBIN'. The accompanying article was lurid and studded with inaccuracies; but it sparked a national furore.

Contrary to accusations in the *Mirror*, the plan in question was not 'secret'. It had been reported for about a year in the nuclear press; but no one outside the industry and a few environmental critics had paid it any attention. The deal immediately at issue was one to reprocess 4000–6000 tonnes of Japanese spent fuel in a new 1000-tonne-per-year oxide reprocessing plant at Windscale. Cost of the plant was now estimated at £300 million. The Japanese would provide down payments and loans of £150 million to help finance construction of the plant; and they would begin delivering spent fuel by 1979. The other half of the plant's capacity would be devoted to reprocessing Britain's own domestic oxide fuel from the AGR stations.

The day the *Mirror* story appeared, Energy Secretary Tony Benn, in answer to a question about it, told the House of Commons that 'the main concern is that the United Kingdom should not become a repository for storing other countries' nuclear waste'. Other aspects of the proposed Japanese contract, and of BNFL's plans for a new oxide reprocessing plant, attracted less concern, either official or unofficial. Benn assured the Commons that any contract would be subject to government consent.

Government control over nuclear activities was soon put in a different and less reassuring light. On 4 November 1975 the Home Office confirmed that the Atomic Energy Authority's own independent constabulary, which guarded nuclear facilities, had access to firearms. Ere long this issue was to come before Parliament, and be acknowledged as a necessary *fait accompli*; but the ostensible responsibility for this third 'armed force' in Britain – separate from either the military or the police – was along a chain of command of transparent tenuousness.

In November 1975 Con Allday, who had taken over as BNFL's chief executive after Norman Franklin's move to the Nuclear Power Company, gave yet another gloss on the Windscale plans. In the company's house newspaper Allday declared that the first new oxide reprocessing plant, with capacity of 1000 tonnes of spent fuel per year, would be in operation by 1983, to handle British spent fuel. The second similar plant would be ready three years later, assuming a government go-ahead for contracts like that with the Japanese. The total investment involved, including additional storage ponds, was now estimated at £900 million.

One of the few Members of Parliament who had paid more than token attention to the activities of BNFL was Robin Cook. He put in a bid for an adjournment debate on the planned Windscale expansion; but he was outflanked by the MP for Windscale, John Cunningham. In consequence the brief debate, on 2 December, consisted of an exchange of laudatory noises about BNFL and its plans, in which doubts scarcely got a mention. Closer to the site in question, *ad hoc* opposition to BNFL's activities at Windscale took the form of protests at the arrival of new shipments of spent fuel at Barrow-in-Furness. In response to local concern BNFL arranged a public meeting in Barrow town hall on 12 December, at which its own senior staff shared the platform with critics and opponents. It was a reasonably well-mannered exchange of views, although it probably changed few minds.

As mentioned in Part I, Energy Secretary Tony Benn was explicitly in favour of public debate on nuclear policy. Prompted by this exercise in a remote corner of Britain, Benn suggested that BNFL might organize a similar get-together at a venue slightly more accessible. It duly took place, on 15 January 1976, in Church House, Westminster, within bowshot of the Abbey and just across the way from the Palace of Westminster. The cast was much the same, except that the meeting was chaired by Sir George Porter, director of the Royal Institution – and that the opening speaker was Benn himself.

On 20 February 1976 it was revealed that BNFL's French partners in United Reprocessors were asking a share in the contract with the Japanese. The Japanese were reported unhappy with the terms and conditions laid down in Britain; Cogema, the French fuel cycle company recently hived off from the Commissariat à l'Enérgie Atomique, began approaching individual Japanese utilities with a more enticing bid. BNFL reacted with indignation, blaming the British government and critics for the loss of half the original order. Critics pointed out that Cogema was BNFL's partner, and would in any case presumably have been entitled to participate. BNFL did not explain; but the relationship between the various members of United Reprocessors appeared to be growing progressively less united.

In the weeks that followed the controversy continued to simmer, but without a real flare-up. Benn, for his part, declared himself satisfied that BNFL's plans were acceptable. On 12 March 1976, in a Parliamentary written reply, he said that he was granting the necessary government approval for the signing of a contract to reprocess 4000 tonnes of Japanese spent fuel at Windscale. This in turn implied approval for construction of the proposed new reprocessing plant to handle the fuel. The Japanese contract was by now stated to be worth £300–500 million – up to half of which might of course go to Cogema. BNFL welcomed the decision, and added that they intended to seek orders to reprocess a total of 6000 tonnes of foreign fuel at Windscale.

Friends of the Earth, who had been leading the opposition to the proposal, decided that since BNFL had come to London to stage a debate, it was only right to return the compliment. In early April FOE chartered a British Rail train for what they called a 'Nuclear Excursion' to Windscale, for a debate on the football pitch outside the Windscale security fence with representatives of BNFL management and staff and Labour and Conservative MPs.

Still pursuing his expressed intention to encourage more open government and more public participation in decisions, Energy Secretary Benn announced a 'National Energy Conference'. The one-day forum took place 22 June, at Church House, Westminster, before an audience of 400. Prime Minister James Callaghan opened the proceedings, followed by fifty-eight speakers of every stripe – chairmen of the nationalized industries and major manufacturers, trades-union leaders, environmentalists and others. Each was rationed to a five-minute stint. Most of the contributions were predictable; but Sir Brian Flowers, chairman of the Royal Commission on Environmental Pollution, used his five minutes to give a tantalizing preview of the commission's long-awaited report *Nuclear Power and the Environment*. What he chose to say dismayed those looking towards an imminent go-ahead for the Windscale plan with no further obstructions. He spoke in measured terms, and afterwards refused to expand or comment; but one passage leapt into the news:

> We believe that nobody should rely for something as basic as energy on a process that produces in quantity a product as dangerous as plutonium . . . We believe that security arrangements adequate for a fully-developed international plutonium economy would have implications for our society which have not so far been taken into account by the government in deciding whether or not to adopt that form of economy.

It was an intriguing foretaste of what was to be a landmark report by the Royal Commission. Nuclear people at once sprang forward to challenge it. They did not, to be sure, challenge the statement itself; Flowers was after all one of them, himself a distinguished nuclear physicist and a part-time board member of the AEA. They could not impute to him the subversive motives they freely imputed to the less eminent among their critics. They did not, however, hesitate, in private and even obliquely in public, to portray Flowers as at best naïve, and at worst a Brutus to the Caesarian cause of nuclear power. *Nuclear Engineering International* for July 1976 headed its editorial comment 'The most honorable Royal Commission chairman of them all?', and invited AEA chairman Sir John Hill to contribute a rebuttal. Hill took the line generally adopted by the nuclear community: 'I am concerned not at what was actually said but at the fact that his remarks have been widely misinterpreted, a misinterpretation that was inevitable in the circumstances.'

This view smacked of wishful thinking; a reading of Flowers's carefully chosen words gives little latitude for 'misinterpretation'. Hill's rebuttal was directed particularly to the impact of Sir Brian's comments as they affected plans for the fast breeder reactor – to which we shall return in Part III. Flowers's monition about a 'plutonium economy' also had, however, a more immediate relevance. A key function of the planned new reprocessing plant at Windscale was, of course, to further just such an incipient 'plutonium economy', separating the material for subsequent use, and presumably shipping it not only around Britain itself but around the world. The point was not lost on opponents of the Windscale plan.

Nuclear controversy in July and August 1976 concentrated on the traumatic abandonment of the steam-generating heavy-water reactor, as described in Part I. On 8 September BNFL announced with great fanfare that it had raised a loan of £100 million from a consortium of twenty-six private banking organizations led by the merchant banking arm of National Westminster. BNFL chairman Sir John Hill declared triumphantly that the loan demonstrated the soundness of BNFL's investment programme, and the faith it inspired on the part of the commercial financial sector. The demonstration of private faith would have been more convincing had the loan not been backed by a 100 per cent guarantee from the British government.

It was perhaps as well for BNFL that the loan was confirmed in early September. Later that month the long-awaited Royal Commission time-bomb at last went off; on 22 September its sixth report, *Nuclear Power and the Environment*, was published. It reverberated through the media, who focused on its uncompromising critique of plans for plutonium fuel and fast breeder reactors. The report was if anything even more emphatic than commission chairman Sir Brian Flowers had been three months to the day earlier. The British nuclear establishment attempted to draw encouragement from the report's dismissal of fears about radiological safety, and about the adequacy of British radiation standards. But the report's strictures on plutonium were coupled with criticism of the lack of a satisfactory plan for final disposal of radioactive waste. The Royal Commission report –

immediately labelled the Flowers report – dramatically raised the temperature of the Windscale reprocessing issue. The report was the first authoritative official acknowledgement that at least some nuclear opposition was based not on far-fetched chimeras but on real and urgent problems – problems the nuclear community had yet to resolve, or even admit.

A week after the report was published, and while its import was still being hotly debated, Cumbria County Council organized a public meeting in the civic hall at Whitehaven, a coastal town some five miles north of the Windscale site. The council was the planning authority that would have to approve BNFL's application for construction of new reprocessing facilities at Windscale. The council's planning committee, which was usually concerned with applications for new garden sheds and the like, had to consider the BNFL application and make its recommendation to the full council. This would be the final decision on the matter, unless the Secretary of State for the Environment, Peter Shore, exercised his right to call the application in for examination by central government. Shore had given no sign that he had any such intention. He appeared prepared to leave the entire £900 million question, with its economic, environmental and diplomatic ramifications, to be decided by the planning committee of the local authority.

The Whitehaven meeting was packed and tense. BNFL speakers warned that refusal to accept a £245 million plan to install new reprocessing facilities might affect the operation of Britain's existing Magnox stations. This allegation clearly referred to the Magnox reprocessing plant, not to the proposed plant for reprocessing oxide fuel; but BNFL appeared to roll all the different activities together – even including those directed to developing 'vitrification' to turn high-level liquid waste into glass for long-term storage or disposal. Opponents too made little distinction between the different categories of activity at Windscale, an unfortunate prefiguration of the confusion that was to continue.

In late October, Friends of the Earth, the Council for the Protection of Rural England and the National Council for Civil Liberties published a study that explored in detail one of the issues highlighted by the Flowers report. Entitled *Nuclear Prospects*, the study analysed the possible impact of civil nuclear activities on civil rights; one of its foci was the need for adequate security to ensure that plutonium – nuclear-weapons material – did not fall into the wrong hands. Providing such security already entailed arming a 'civil' nuclear police force; it might also require vetting and surveillance of individuals, secret dossiers on private citizens, and other serious infringements of commonly accepted rights in a free society. Actual theft of plutonium, with the threat of nuclear blackmail, would almost certainly be followed by the equivalent of martial law. The report was acknowledged as dispassionate, well argued and profoundly disturbing; *The Times* devoted a lengthy leader to its findings and implications, not least those for the Windscale reprocessing plans.

Nor was such criticism confined to Britain itself. On 28 October 1976, the week before the US presidential election, President Gerald Ford delivered a statement dramatically revising US civil nuclear policy. He began by noting the importance of nuclear power as an energy source; then he continued,

> On the other hand, nuclear fuel as it produces power also produces plutonium, which can be chemically separated from the spent fuel. The plutonium can be recycled and used to generate additional nuclear power, thereby partially offsetting the need for additional energy resources. Unfortunately – and this is the root of the problem – the same plutonium produced in nuclear power plants can, when chemically separated, also be used to make nuclear explosives . . . Developing the enormous benefits of nuclear energy while simultaneously developing the means to prevent proliferation [of nuclear weapons] is one of the major challenges facing all nations of the world today.

Ford went on to declare flatly:

> I have concluded that the reprocessing and recycling of plutonium should not proceed unless there is sound reason to conclude that the world community can effectively overcome the associated risks of proliferation . . . I have decided that the United States should no longer regard reprocessing of used nuclear fuel to produce plutonium as a necessary and inevitable step in the nuclear fuel cycle.

Ford then explored the national and international implications of this decision. He called on other nations to join with the US in confronting the problem of proliferation and its link to civil nuclear power, including reprocessing and the plutonium-fuelled fast breeder reactor.

A week later Ford was out of the White House, making way for incoming President Jimmy Carter. Carter's election did not, however, ease the international tension created by plans for 'civil' plutonium: on the contrary. Carter's election platform had laid great stress on the problem of civil nuclear power and nuclear weapons proliferation; and his administration would soon be at loggerheads with many otherwise friendly governments elsewhere – not least the British government.

The British government meanwhile was at last having to contend with a rising tide of unease about the Windscale plans. Even those who were not fundamentally unsympathetic to the idea of reprocessing or the import of foreign spent fuel were none the less unhappy that the ultimate decision was to be left to the planning committee of a single small local authority. The Secretary of State for the Environment had recently 'called in' an application to expand the town dump at Penrith, north-east of Windscale; surely it would be absurd in the circumstances to leave unexamined a 'waste management' decision of so much greater magnitude? Government spokespeople pointed out that Energy Secretary Benn had given the official go-ahead to the general proposal the preceding March; it would be improper to reopen the issue retrospectively. Be that as it might, the clamour continued to mount. Among those insisting that Shore call in the application were Friends of the Earth, the Council for the Protection of Rural England, the Town and Country Planning Association, the Conservation Society, the Socialist Environment and Resources Association, and the Lawyers' Ecology Group.

On 2 November the Cumbria County Council planning committee announced that 'subject to appropriate conditions' it was 'minded to approve' BNFL's application; but that it was a departure from the county structure plan – thus putting it back into Environment Secretary Shore's in-tray. On 3 November Shore told the Commons that the Windscale proposals would accordingly come before him after all; he agreed to report about them to Parliament. Some MPs urged him to call in for an inquiry the part of the application relating to oxide fuel reprocessing; Shore gave no indication that he might do so.

Shore had twenty-one days to approve or reject the 'change to the structure plan'. If he did nothing it would go through. The day before the deadline the Lawyers' Ecology Group delivered a letter to him, signed by three eminent planning law QCs, warning that if he failed to call in the Windscale application, the Cumbria Council decision might be open to challenge in court, since it was not in compliance with the relevant planning Act. On 25 November, at the very last minute, Shore told the House of Commons – in a written reply not subject to supplementary questions – that 'I am still considering this matter. I have directed the council not to grant permission until I have reached my decision.'

BNFL was disappointed at the delay. Chairman Sir John Hill said:

> We agree that the Government should take all reasonable time to come to its decision, but we are concerned that if the matter is called in for a public inquiry, all this foreign business will disappear – and with it 5000 jobs. Already the French have taken half the business being offered by Japan, because of this year's delays, and we have now reached a situation with our potential customers where they will soon have to take action very soon to cope with their fuel if we cannot give them an answer.

As before Hill gave no explanation of the relationship between BNFL and their French partners Cogema in United Reprocessors, and the way they shared foreign business. Nor did he explain what other 'action' might be taken by 'potential customers'. As it turned out, Hill's hand-waving allegations about the possible consequences of 'delay' bore little relation to practical reality. In the event, more than six years later BNFL would not be ready even to submit a detailed planning application for the new oxide reprocessing plant.

Even in 1976 BNFL had its hands full coping with existing problems at Windscale. At the height of the public furore about the Windscale planning application, Energy Secretary Tony Benn dropped another bombshell. On 9 December Benn revealed to the Commons that BNFL had discovered a leak of radioactivity from a waste storage 'silo' – a vast concrete building – on the Windscale site. Indeed they had discovered it six weeks before, in early October, while Cumbria Council was weighing BNFL's planning application. BNFL had kept the news of the leak secret; word of it had reached Benn only in December. When the story became public, a BNFL representative offered a remarkable explanation for keeping quiet about the leak. BNFL thought it would be imprudent, he said, to release the information about the leak at a time when it could have been interpreted as a manoeuvre to press for planning permission to renovate and expand the factory.

Other observers noted drily that publicity about the leak might more plausibly have scuttled the entire Windscale expansion plan. Certainly Benn himself was outraged at the cover-up. Reliable reports indicated that the Windscale issue was now the subject of heated conflict in the Cabinet. Benn, having given government approval in principle to the plan in March, had hitherto apparently aligned himself with those other Ministers, especially Industry Minister Eric Varley, who wished to grant the go-ahead forthwith. After the revelation of the silo leak and its cover-up, Benn was reported to have changed his stance, and joined his voice to that of Shore in pressing the Cabinet to agree to an inquiry. Shore for his part was said to feel that if he did not call in the Windscale application he would be setting an impossible precedent: if the Windscale decision did not require the imprimatur of central government, no subsequent planning decision could possibly do so.

The leak at Windscale was somewhere in the lower regions of building B38. This building was a massive concrete bunker, used to store the broken fragments of metal cladding that had been stripped off spent Magnox fuel in preparation for dissolving the uranium rod in acid. The magnesium alloy Magnox fragments were themselves heavily contaminated with radioactivity. They were also very chemically reactive, and had to be stored under water to prevent spontaneous combustion. Thus stored, however, they slowly deteriorated into a semi-solid radioactive sludge on the floor of the storage silo; and the covering water leached out their radioactivity. It was this radioactive covering water that was now seeping into the ground below B38, at about 400 litres a day. The leak had been discovered by accident, during excavations near the silo. BNFL declared that the leaking radioactivity presented no hazard to site staff or to the public. Unfortunately, however, it was in due course to become apparent that neither BNFL nor anyone else actually knew how to stop the leak.

On 22 December, following one final discussion in Cabinet, Peter Shore announced in the House of Commons that there would, after all, be a public inquiry into the Windscale proposals. Shore invited BNFL to disentangle the three strands of its expansion plans, and submit three separate planning applications. 'If and when a separate application is submitted for the oxide fuel reprocessing plant, I would call it in to deal with myself and order a public inquiry. This will enable all the relevant safety, environmental and planning considerations fully to be examined.'

BNFL chairman Con Allday reacted angrily, calling the inquiry 'unnecessary', and insisting that a full debate had already been held during the past year. If the inquiry were to drag on, it would damage the credibility of BNFL abroad; 'If overseas business is lost, it will be damaging to this country as well.' Allday did not apparently feel any need to add that the only civil reprocessing plant of larger than pilot scale actually in operation anywhere in the world was the Haute Activite Oxide head end plant of Cogema at Cap la Hague on the north French coast. Not only was this plant already fully committed to reprocessing fuel from France's own domestic nuclear programme; it also belonged to BNFL's own partner in United Reprocessors. To whom was BNFL going to 'lose' the putative overseas business? As it was to turn out, BNFL did indeed 'lose' business – but not to other reprocessors; and the long debate over the Windscale proposals had nothing whatever to do with the 'loss' of business.

On 27 January 1977 thirty-two changing-room attendants at Windscale walked out in a dispute over working conditions; some 3000 manual staff had to be sent home. When the manual workers found out that the company was not proposing to pay them for the time lost, they too walked out. Two of the remaining three operating Calder Hall reactors had to be shut down, as did the Windscale reprocessing plant. Within a few days some 3000 Windscale workers were on strike, and another 2000 in enforced idleness. After the strike had lasted a month, reports began to circulate that the site was facing safety problems because of a shortage of electrical power. On 6 March pickets stopped a load of carbon dioxide from entering the Windscale gates; it was admitted only after Energy Secretary Tony Benn sent a telegram to the strikers asking them not to do anything to endanger safety at Windscale. The following day pickets blocked a tankerload of nitrogen needed to replenish the inert cover-gas in plutonium facilities. A union representative declared that: 'The quickest way to remove the hazard is by immediate negotiation with a view to resolving the strike.' By 11 March reports said that troops might have to be called out to take essential safety supplies on to the

Windscale site. The next day Benn himself flew to Windscale to talk to the strikers; but he failed to persuade them to let the supplies through. At length, after forty-eight hours of intensive negotiation, BNFL made an improved offer and the strikers voted to accept it, ending the stoppage. The union's Windscale convenor suggested that blocking the nitrogen shipment had been the 'trump card' to force Benn to intervene and BNFL to raise its offer. It was not exactly a reassuring omen for the future of the installation.

While BNFL management was attempting to get its existing facilities back to work, it was also submitting new planning applications for its proposed expansion, separated into not three but four parts. On 1 March they came before Cumbria County Council planning committee, which forthwith granted outline planning permission for 'refurbishment' of the Magnox reprocessing plant, and for development of the so-called 'Harvest' process for turning high-level liquid waste into glass. The other two applications were for the provision of new storage ponds for oxide fuel, and for the proposed new oxide fuel reprocessing plant itself. These two applications were forwarded to the Department of the Environment. Observers noted that BNFL might have been hoping that Environment Secretary Shore would 'call in' for an inquiry only that application relating specifically to the reprocessing plant itself, and let the ponds go ahead. In that way the company could assure its prospective customers that it could accept their spent fuel, even if it could not guarantee to do anything with it except store it. Events were to suggest that customers were indeed eager simply to have their spent fuel taken off their hands.

On 7 March, as expected, Shore announced that there would be an inquiry into the application for the proposed Thermal Oxide Reprocessing Plant – thereafter labelled THORP. The inquiry was to be held in Whitehaven civic hall, commencing on 14 June, and would be chaired by Roger Parker, a High Court justice best known for chairing the inquiry into the Flixborough chemical plant explosion. The imminence of the inquiry threw objectors into a frenzy of fund-raising and organization, preparing their cases and seeking legal assistance. In due course, despite the company's insistence on the urgency of the matter, it was found also to have caught BNFL far from fully prepared.

On 7 April US President Jimmy Carter picked up the theme raised by his precursor Gerald Ford the preceding October. Carter issued a policy statement withdrawing US government support for reprocessing, plutonium fuel and the fast breeder. The original intentions of those drafting the statement had been to send an international signal, to set an American example that might help to persuade Britain, France, Japan and other countries to think again about their plans for reprocessing and use of plutonium. At the press conference introducing the statement, however, Carter interpolated certain off-the-cuff glosses that severely attenuated the international impact of the policy statement. In Britain, BNFL and the British government seized on these comments and declared emphatically that the new US policy on plutonium had therefore no bearing on plans for Windscale or British fast breeder reactors.

Carter's policy statement nevertheless had concealed teeth, of which his foreign critics in government and the nuclear industry were well aware. Almost all the fuel used in power reactors in Western industrial countries other than Britain, France and Canada contained uranium that had been enriched in the US. The contracts for such enrichment provided that the US government had to give permission for all subsequent transportation and processing of the resulting spent fuel. For many years this permission had been granted essentially automatically and without question. Now, however, the Carter administration let it be known that it would be in no hurry to authorize the transfer of US-enriched spent fuel from a foreign client country to another for reprocessing. In particular the US could invoke its contractual rights to forbid Japanese electricity companies from shipping their spent fuel to Windscale, or indeed to the French sister-plant at Cap la Hague. Both the British and French governments pronounced themselves unperturbed by the new US policy and its implications for British and French nuclear activities. Behind the scenes, however, a deep division was opening between the US and almost every other government in the nuclear-industrial West. It was to lead to intense and prolonged top-level diplomatic controversy. The international plutonium controversy did not, however, seriously impede BNFL's plans. They were, to be sure, impeded – but not by Jimmy Carter.*

* See *The Plutonium Business*, Walter C. Patterson, Paladin, 1984.

On the weekend of 13–14 May, Energy Secretary Tony Benn presided over a discreet seminar on British nuclear policy. It was held at the Civil Service College at Sunningdale, west of London, and attended by almost all the top brass of the electronuclear policy establishment – Hill, Marshall, Allday, Franklin, Tombs, Cottrell, Flowers, Sir Herman Bondi, John Dunster of the Health and Safety Executive, several Select Committee MPs, civil servants and two nuclear dissenters – Sir Kelvin Spencer and the present author. A concise summary of the proceedings of the Sunningdale seminar was later published; the divergence of views between the electronuclear people and the doubters was comprehensive.

On 14 June 1974 the Windscale inquiry opened, in the civic hall at Whitehaven, five miles up the west Cumbrian coast from Windscale itself. By the time it rose for the last time, on 4 November, it had been sitting for exactly 100 days. Every conceivable argument for and against the THORP proposal had been canvassed and challenged; some arguments had been retraced so many times that they had long since fallen into a sort of call-and-response litany like a Latin mass:

BNFL: THORP would be technically straightforward, a simple extrapolation from existing reprocessing plants and BNFL's experience.

OBJECTORS: No one anywhere had built and operated a successful commercial oxide reprocessing plant. Those who had tried, in the US, Belgium, Federal Germany and France, had met with endless trouble; and BNFL's own experience, with the B204 Head End Plant, was hardly reassuring.

BNFL: THORP would be necessary to process spent oxide fuel for final disposal.

OBJECTORS: No one yet knew what arrangement would eventually be made for final disposal; until this was agreed it made more sense to keep all options open. Chopping up and dissolving intact oxide fuel destroyed the option of storing it intact; furthermore it created a whole range of new radioactive wastes, solid, liquid and gaseous, some of which were discharged directly into the air and water or dumped on land, and some of which were far more difficult to store than the original intact spent fuel.

BNFL: THORP would recover valuable uranium and plutonium for use in making more nuclear fuel.

OBJECTORS: The recovered uranium would cost far more than the abundant fresh uranium coming on to the market around the world; the plutonium would be useless and valueless until there was a major programme of fast breeders to burn it; and in any case plutonium from oxide fuel was much poorer in fuel quality than that from Magnox reactors.

BNFL: It was preferable to reprocess spent fuel from foreign countries at Windscale, than to have the foreign countries themselves acquire reprocessing plants that could recover plutonium for weapons.

OBJECTORS: Not if BNFL delivers the separated plutonium back to the foreign countries; in any case Japan, for instance, is already about to start up its own reprocessing plant at Tokai Mura. If Britain, with all its other energy resources, insists on the need to embark immediately on the use of plutonium for fuel, any other country can reasonably claim a similar need, even if the other country, like Pakistan or Argentina, might have a hidden alternative use for separated plutonium – nuclear weapons. It would be far better for Britain to follow the example of the Carter administration in the US, and acknowledge that there is no need to use plutonium for any civil purpose.

All these and many other arguments were rehearsed and rehashed until the few remaining onlookers in the civic hall were glassy-eyed. When the inquiry adjourned for the last time no tears were shed on either side. Meanwhile, as the inquiry was still in its dying stages, the Atomic Energy Authority, Friends of the Earth and other interested organizations arranged to co-sponsor a two-day conference at the Royal Institution in Mayfair, under the title of 'Nuclear Power and the Energy Future'. Leading figures from the nuclear establishment and its critics joined in a series of six sessions of debate before an open audience of some 400 people, drawn from both the nuclear community and the general public. The disputation was intense, occasionally verging on the acrimonious; but it was nevertheless an impressive demonstration of the possibility of 'rational discussion' of nuclear issues, the ideal approach so often lauded by nuclear spokespeople. It was probably the high-water mark of dialogue between nuclear proponents and critics in Britain. Within less than six months Mr Justice Parker was to publish his official report on the Windscale inquiry; from that day onwards, even those critics who had looked for dialogue and rational discussion decided that official nuclear policy was beyond influence by rational argument.

In the interlude between the end of the inquiry and the publication of Parker's report, it became abundantly apparent that the existing procedural mechanisms for handling such an issue fell far short of adequacy. When a routine planning application was called in for consideration by the Environment Secretary, the Secretary would publish the inquiry report only on the day he announced whether the application was being accepted or rejected; the report was published only to provide background information as to the decision reached. In the weeks after the inquiry ended, Shore let it be known that he saw no reason to vary this arrangement for the Windscale decision. Many of his Parliamentary colleagues thought differently, as did objectors like Friends of the Earth. They mounted a campaign calling for publication and discussion of the inquiry report before any political decision was taken. In due course over 200 MPs signed a so-called Early Day Motion pressing Shore to publish the report before taking his decision.

Parker submitted his report to Shore in late January. By that time the weight of opinion, not only in Parliament but in the editorial columns of the national press, was comprehensively aligned behind the call to publish the report before announcing the official decision. At last, on 6 March 1978, the report was published. Security was stringent; no copies were made available to press or objectors beforehand. One journalist with inside connections to the nuclear industry had however given an accurate foretaste a few days earlier: not only had Parker found in favour of BNFL and THORP, he had dismissed every argument of the objectors out of hand.

At 3.30 P.M., 6 March, Shore rose in the House of Commons to introduce the report, and to declare that he had found it completely persuasive. Nevertheless, he accepted the force of the argument calling for Parliamentary debate before reaching a decision. Accordingly, since the normal planning mechanisms made no provision for this, he proposed to adopt an extraordinary procedure to handle the issue. He was formally rejecting the BNFL application; there would then be a House debate on the Parker report, on 22 March. Shore would then lay a so-called Special Development Order, authorizing construction of the Thermal Oxide Reprocessing Plant (THORP) after all. This Special Development Order would in turn be debated on 15 May, and submitted to a vote of the House for approval.

No one could have called the arrangement tidy; but to poleaxed objectors its surreal convolution seemed somehow entirely appropriate. Those like Friends of the Earth who had come away from the inquiry itself feeling that they had made a strong case, and that Parker had taken it on board, could find in the ninety-nine terse pages of the Parker report no sign of their case whatever. Their arguments had not been refuted so much as simply ignored. Parker gave no indication why he ignored them, nor why he accepted points put forward by BNFL that had apparently been refuted in cross-examination. BNFL was of course exultant. Company spokespeople and their supporters made much of the discomfiture of the objectors, and the unconditional endorsement Parker gave to THORP. As matters were to turn out, BNFL was well advised to exult while it could. From 6 March 1978 onwards the fortunes of the reprocessors were to be downhill all the way.

8 Gone with the Windscale

The Windscale inquiry, and the events that were to follow, put Windscale so egregiously on the map that BNFL eventually tried to take it off again. On 22 March 1978 the House of Commons duly assembled to debate the Parker report. The small group of Liberal MPs decided to insist on a division. Shell-shocked objectors doubted the desirability of a vote at that stage, when the impact of the Parker report had yet to be countered by rebuttals. Nevertheless, when the House divided, as well as 186 votes in favour of THORP, there were 56 against – much the largest vote ever recorded in opposition to any civil nuclear decision in Britain.

On 15 May, in the debate on the Windscale Special Development Order, the government imposed a two-line whip to bring out its backbenchers and junior Ministers. Even so the final vote was 224 to 80; and the 80 MPs voting against the Windscale plan ranged across the entire breadth of the political spectrum. It is difficult to imagine many other issues that could prompt, for example, both John Biffen MP and Robin Cook MP into the same voting lobby; but they both voted against the Windscale order. To be sure, the opposing votes made no practical difference: THORP had received the go-ahead, sought for so long and with such urgency by BNFL. Once given the go-ahead, BNFL might have been expected to go ahead: to embark on the construction of the plant they had represented as immediately essential for the future of British nuclear power. It did not work out quite like that.

On 24 May, only two days after the Commons debate, BNFL's chairman and managing director, Sir John Hill and Con Allday, sat down in Tokyo with representatives from the ten Japanese electricity companies and signed contracts said to be worth nearly £300 million for the reprocessing of 1600 tonnes of Japanese spent fuel, and another £200 million for transporting it to Windscale from Japan. The transport contract also covered another 1600 tonnes of spent fuel, to be taken to the French reprocessing plant at Cap la Hague. Senior executives from BNFL's French partners Cogema were also present; they had signed in September 1977 a contract for reprocessing this second 1600 tonnes at la Hague. The spent fuel was to be delivered between 1982 and 1990. BNFL was reported to be negotiating contracts with electricity companies in Europe, aiming to book a total of 6000 tonnes of fuel from British and foreign customers, to fill what it now called the nominal ten-year operating capacity of THORP. That, to be sure, worked out at 600 tonnes per year, as against the 1200-tonne-per-year capacity claimed for THORP at the Windscale inquiry. Observers were by this time, however, aware that BNFL, like reprocessors everywhere, described the capacity of a reprocessing plant the way the children's riddle described the length of a piece of string. The capacity of a reprocessing plant was twice the capacity of half the plant. The effect of this elastic concept on the unit capital cost of reprocessing a tonne of fuel would have been more striking if the Japanese had not agreed to a contract charging them cost-plus: they were to pay – much of it in advance – whatever BNFL told them the plant cost, plus a profit for BNFL. BNFL was to find that not all its prospective customers were that eager to get shot of their spent fuel. Among the less eager were to be the CEGB and the South of Scotland Electricity Board, on whose alleged behalf the entire enterprise was being undertaken.

Nor was the question of oxide fuel the only one to give rise to domestic awkwardness in Britain itself. BNFL's long-running troubles with Magnox fuel reprocessing, described in an earlier chapter, were to be tackled by building a complete replacement for the Magnox storage ponds and decanning plant, on a different part of the Windscale site. This so-called 'refurbishment' entailed an investment of at least £245 million. Such was the estimate BNFL was using while promoting the Nuclear Industry Finance Bill in 1976; like many another nuclear estimate this cost thereafter mounted steadily. Somebody had to pay it; and BNFL was insistent that the somebody should be the CEGB, by paying higher charges for having its Magnox fuel reprocessed in the new facility. The CEGB, however, was disinclined to accept BNFL's reasoning. The CEGB had a long-term contract with BNFL for Magnox fuel reprocessing, on terms that were by the late 1970s – after years of double-digit inflation – distinctly advantageous. The CEGB was also by no means kindly disposed towards BNFL, as a result of BNFL's unilateral refusal to accept shipments of CEGB Magnox fuel during the height of the Magnox reprocessing troubles in the mid-1970s.

At Wylfa, the last and largest of the Magnox stations, spent fuel was already being discharged not into a water-filled cooling pond but into a storage magazine cooled by carbon dioxide gas. This dry-storage system dramatically reduced the problem of corrosion of spent Magnox fuel; it was after all designed to operate in a carbon dioxide atmosphere in the reactor itself. The Wylfa dry-storage facility cooled by carbon dioxide proved so successful that the CEGB decided to add two further units, cooled by natural circulation of ordinary air. It also participated in studies carried out by the facility's manufacturers, GEC Energy Systems, to evaluate the concept for dry storage of spent oxide fuel. The CEGB was clearly not party to BNFL's vehement insistence at the Windscale inquiry that dry storage, as strongly advocated by Friends of the Earth, would be difficult, expensive and dangerous. On the contrary: within five years the CEGB would itself be lining up with those favouring dry storage of spent fuel. In so doing it would be putting yet further pressure on BNFL's reprocessing aspirations.

In October 1978 the CEGB and other interested parties were reminded of the potentially precarious state of affairs at Windscale. By that time almost everyone except the site staff had forgotten about the stubborn leak under storage silo B38; but the storage silos had fresh surprises in store. One of the effects of radiation on water is to break up its molecules, releasing so-called 'radiolytic hydrogen'. Production of this hydrogen has to be carefully monitored and controlled; under appropriate conditions hydrogen is a powerful explosive, and accordingly undesirable in a radioactive waste-storage facility.

On 31 October BNFL staff at Windscale discovered that hydrogen had begun to accumulate to an alarming level in the currently operating Windscale storage silo for radioactive – and inflammable – Magnox cladding scrap. Personnel were evacuated; since the silo could accept no more Magnox scrap, the B30 decanning unit and the

B205 reprocessing plant had to be shut down. No one knew why the silo had suddenly begun to accumulate hydrogen. Site staff suspected an unusual subsidence of the stored Magnox scrap; but until the Nuclear Installations Inspectorate carried out an investigation to determine, and presumably eliminate, the cause, the plant had to remain shut down. The incident was only one of a succession of cumulative embarrassments for BNFL at Windscale. Observers recalled that the Flowers report had included some strictures about the doubtful 'housekeeping' at Windscale. As the months passed and the incidents multiplied the criticism seemed all too apposite.

The B38 bunker continued to leak. Radioactivity levels in the soil beside the bunker gradually increased, until they became too high to permit further excavation; the existing hole in the ground was marked off-limits and the work suspended. Other BNFL staff, however, carried on drilling exploratory holes to sample the subterranean radioactivity profile of the site. On 15 March 1979 they got a nasty surprise. In one of their boreholes they found radioactive material that could not have come from the leak under the B38 bunker. In consternation they realized that it was in fact high-level liquid waste. Hurried investigations tracked the source of this radioactivity to a small disused building known as B701, an adjunct to the storage-tank building B215. Those with long memories recalled that B701 bore the remarkable title of Export Plant. In it was a spur from the main pipeline carrying high-level waste from the chemical separation plant to the storage tanks in B215. This spur led to a valve through which samples of the high-level waste could be tapped off into containers for transport to Harwell, to be used in Harwell's work on encapsulation of such wastes in glass. The Harwell work had however been terminated in the mid-1960s, for reasons that have never been clearly explained. B701 thereafter appeared to have been virtually forgotten.

Left to itself, its plumbing at some stage began to leak. High-level waste spilled on the floor, gathering in the sump below the take-off point. The sump overflowed, and the room itself began to fill; in due course high-level waste liquid began to seep into the soil below and around B701. By the time the leak was at last discovered it had been leaking for many years; an estimated 10,000 litres of liquid containing some 30,000 curies of radioactivity had escaped into the soil. By mid-April 1979 the leak had been located and stopped; but the episode further undermined confidence in the quality of BNFL's housekeeping at Windscale.

In mid-July 1979 Windscale was back in the news with more egg on its face. A fire broke out inside the shielded 'cave' in building B30 where Magnox fuel was stripped of its cladding. The fire was not serious, although it slightly contaminated eight workers and necessitated a plant shut-down for several days. But it further added to the burden of background radiation in B30; B30 and the old cooling ponds were becoming progressively less salutary places to be, requiring in some areas the wearing of full protective gear.

In the spring of 1979, even before the B30 fire, the Nuclear Installations Inspectorate had told the government that it was time for a coordinated investigation of safety and management at Windscale. After the general election the incoming Conservatives agreed to the suggestion. An NII team set to work in September 1979 to study managerial arrangements for safety, review procedures for safety surveillance in operating units at the site, and examine how the safety of plant was assessed. The stated intention was to have the report completed by the end of 1979, and presumably published shortly thereafter. The report was not in fact published until February 1981; the delay was apparently occasioned by the NII staff shortage mentioned in Part I, and by what the NII investigating team found when they set to work at Windscale.

By that time the NII had already published painstaking reports on two of the more noteworthy episodes of recent Windscale history. One concerned the leak under the B38 storage silo. The report, published in February 1980, declared in its introduction that 'The investigations have reached the stage where a number of possible courses of action have been identified which might alleviate the problem of the leak and deal with the activity which has leaked out. This may involve work in high radiation fields but will certainly be time-consuming and costly, and will involve the development of special equipment and techniques.' It then went on to spell out just why it reached these discouraging conclusions. Anyone reading the report would understand immediately why BNFL was less than keen to embark on any attempt to rectify the mess under B38.

The other NII report concerned the leak of high-level waste from the so-called Export Plant, discovered inadvertently during the detective work around and below B38. As described above, the plant had been used to tap off liquid high-level waste for shipment to Harwell for work on converting it into glass. According to the report, 'the last such consignment was sent in 1958. There is no evidence that the building has ever been decommissioned. Because of the possibility of spillage the inside walls and floor . . . are clad with metal and drain into a sump vessel. Owing to high radiation levels there is no access by personnel.' It then revealed that 'video-camera examination showed liquid up to 10–15cm from the top of the metal cladding'. The investigators found that even with plumbing correctly aligned 'radioactive liquor could splash over into the Export Plant aging tank. This tank then overflowed into the sump vessel. Because operating instructions for emptying the sump vessel . . . were not complied with, the radioactive liquor eventually filled the sump vessel and overflowed into the metal clad area in the bottom of building B701.' It then 'escaped through defects in the metal cladding and finally leaked into the ground . . .' Although the original stories about the leak referred to an estimated loss of 30,000 curies of radioactivity, the NII team put the figure at 'rather more than 100,000 curies'. The report concluded that 'the operational system was not adequate to maintain control over radioactive liquors. Despite certain design weaknesses, the engineered part of the system . . . would have been adequate to prevent the incident occurring if the operating instructions . . . had been carried out'.

By the time the overall NII report on 'The Management of Safety at Windscale' appeared, in February 1981, there was thus already a substantial body of public information pointing to its findings. The report took pains to stress that 'we would not like . . . to give the impression that we regard Windscale as a dangerous place to work or near which to live'. Nevertheless, the report put forward a lengthy catalogue of conclusions and recommendations worded in uncompromising terms. The first finding set the tone: 'By the early 1970s the standards of the plant at Windscale had deteriorated to an unsatisfactory level. We consider this represented a poor base line from which to develop high standards of safety. We are strongly of the opinion that such a situation should not have been allowed to develop, nor should it be permitted to occur again.' Other findings included the following:

We found that instructions as to the full extent of ... responsibilities [of Works Managers] were not always clear and comprehensive and we were not convinced they were clearly understood by Works Managers and their staff, or were being adequately performed ...

A few incidents, including the two major leakages of radioactivity into the ground, have been a cause for concern to us because of the implications of multiple failures of safety precautions. There is evidence of a failure to learn from previous events which should have been recognized as indications that these incidents might occur ...

It was clear to us that insufficient attention has been given to extending plant operating instructions to deal with reasonably foreseeable abnormal plant operating conditions.

The final finding asserted bravely that:

It will never be possible to eliminate entirely the occurrence of incidents, especially those in which human error or poor judgement play a contributory part. There can be no absolute assurances that incidents of the same kind as have previously been reported from Windscale will not occur in the future. Nevertheless we believe that the rate of occurrence and the potential consequences can be reduced by careful adherence to well-conceived safety precautions, and in particular by careful attention to the preparation of, strict compliance with, and regular review of safety procedures.

The nuclear inspectors should have kept their fingers crossed.

In May 1981 BNFL reorganized its management structure – and made one other interesting change. It announced that henceforth its site on the west Cumbrian coast would be known as Sellafield, not Windscale. Although BNFL stoutly denied any such implication, many observers concluded that BNFL was by this time finding the popular image of 'Windscale' burdensome. If such had indeed been the case, events were ere long to make 'Sellafield' a name almost as notorious as Windscale. In any event, the reprocessing facilities continued to be called the Windscale Works; and many press and broadcast commentators continued to use the name. Those who accepted 'Sellafield' tended to add 'formerly known as Windscale', defeating any subliminal BNFL hope that the change of name would bring with it a change in popular perceptions.

By midsummer 1981 work was well underway on construction of the new Pond 5 reception and decanning unit and other Magnox reprocessing facilities at Windscale – the so-called 'refurbishment' approved in 1977. The cost, estimated at £245 million in 1976, was by 1981 being stated as some £850 million at January 1980 prices. The Site Ion Exchange Effluent Plant (SIXEP) was likewise under construction, to reduce the radioactive discharges into the Irish Sea. Together with a plant for medium-active waste it was expected to cost around £100 million. The Thermal Oxide Reprocessing Plant (THORP), given the go-ahead in 1978, was just a hole in the ground, although it was still expected to be in operation by the late 1980s. Its design capacity was still being stated as 1200 tonnes of uranium per year; its capital cost was by this time estimated to be £800 to £1000 million – a range of possibility suggesting that actual costing was as yet at a very preliminary stage, despite the forthright estimates BNFL had been prepared to quote as far back as 1974. Capital investment on other reprocessing support facilities, and their necessary design engineering, brought BNFL's anticipated total outlay on reprocessing at Sellafield to £2000 million over the coming decade.

This expenditure was, of course, predicated on the assumption that customers, domestic and foreign, would want to have their fuel reprocessed by BNFL at Sellafield, and would be prepared to pay the going rate. By 1981, however, the faintest of shadows had begun to lengthen over this confident assumption. The August 1981 issue of *Nuclear Engineering International*, containing the article outlining BNFL's corporate plans, had on its cover a diagram of a facility designed by GEC Energy Systems, for the long-term dry storage of oxide fuel. Details were provided in a long and enthusiastic article by GEC engineers. At the Windscale inquiry four years earlier, Friends of the Earth had advocated development of this technology as a preferable alternative to reprocessing. BNFL had decried the idea as too expensive, too difficult and too dangerous. By 1981, nevertheless, it was receiving increasingly enthusiastic support from electricity suppliers in several other countries; and the installations already operating were proving to be comparatively cheap, straightforward and reliable.

Added to this was the dawning realization that the postulated shortage of uranium, which would render imperative the use of plutonium-based fuels, was a mirage. On the contrary: uranium mining companies were finding to their horror that there was so much uranium coming on to the market, with new mines opening amid a drastic cutback in nuclear programmes everywhere, that the price of uranium was falling steadily. From a high of over $40 a pound of yellowcake it had dropped by 1981 to less than $20, and showed no sign of rising significantly in the foreseeable future.

Taken together, the feasibility of long-term storage of spent oxide fuel and the ready availability of fresh uranium set an upper limit of sorts on the prices that electricity suppliers might be prepared to pay to have their spent oxide fuel reprocessed. If instead they could simply store it, and buy fresh uranium for fuel at a better price than would be charged for fuel made from reprocessed uranium and plutonium, why should they even bother with reprocessing? Some such reasoning had clearly begun to percolate even within the precincts of the CEGB. Unlike the desperate Japanese in the late 1970s, the CEGB declared itself unprepared to write BNFL a blank cost-plus cheque for the reprocessing of oxide fuel. To be sure the CEGB had 'reserved' half the capacity of THORP for CEGB fuel; but as the months and years rolled by, the CEGB and BNFL remained locked in stalemate over terms, and the contract remained unsigned.

On 4 October 1981 instruments at Sellafield detected the radioisotope iodine 131 being emitted from the reprocessing plant. Iodine 131 is one of the most biologically dangerous radioisotopes produced during a chain reaction; but it has a 'half-life' of only eight days, and decays so rapidly that it has essentially disappeared before spent fuel is reprocessed. The plant was shut down, while urgent investigations began into the source of the iodine. Following the directive by Tony Benn during his stint as Energy Secretary, BNFL had been for several years scrupulously reporting the most trivial incidents at Sellafield. This time, however, only after several days of cumulatively more circumstantial rumour did BNFL admit officially that iodine had been released. Subsequent inquiries established that the iodine had come from six fuel elements freshly discharged from the Hinkley Point Magnox station, that had been shipped to Sellafield in error and reprocessed prematurely; the amount of iodine released was found to be insignificant. The incident could have been written off as a minor administrative hiccup – were it not that BNFL, faced with a sudden and unexpected problem that might have been serious, had lapsed again into its long-standing habit of secrecy.

Throughout most of 1983 nuclear policy interest in Britain was focused on the Sizewell inquiry, as described in Part I. On 30 October, however, BNFL and Windscale a.k.a. Sellafield were back on the front pages yet again. The initial impetus this time was a programme from Yorkshire TV called 'Windscale: the Nuclear Laundry'. The programme alleged that the incidence of cancers, particularly childhood cancers like leukaemia, in the vicinity of Windscale was much higher than the national average. It also reported that household dust in a town near Windscale on the west Cumbrian coast was found to contain plutonium. The programme triggered a national uproar, so much so that the government asked Sir Douglas Black, a past president of the Royal College of Physicians, to chair an inquiry into the cancer incidence in the area and its possible causes. The Black inquiry in due course reported in curiously ambivalent terms, declaring apparently that there was no reason to worry about the Sellafield site but that there might be. Few found the report entirely satisfactory.

While this uproar was at its height, the environmental organization Greenpeace sent a boat to the Irish Sea off the coast near Sellafield, declaring its intention of blocking the effluent pipeline out of which the plant poured half a million gallons of slightly radioactive liquid waste daily. BNFL countered by sending its own divers down covertly to alter the configuration of the pipe outfall so that the Greenpeace plug would not fit. BNFL also sought and obtained a High Court injunction to prevent Greenpeace from tampering with the outfall. Greenpeace persisted, and was slapped with a fine initially set at £50,000.

By that time, however, people all over the country were asking who the real transgressors were: because a team of Greenpeace divers had emerged from the water extensively contaminated with radioactivity far more concentrated than that which Sellafield was supposed to be discharging. A hasty investigation, followed by a much more extensive one, revealed that the Greenpeace divers had been in the water while Sellafield staff were trying to cope with an internal cock-up – one that probably would have gone unreported had Greenpeace not been there. A staff oversight had diverted radioactive process liquid into the wrong vessel. Once there it was impossible to remove; the staff accordingly decided just to discharge it out of the pipeline with the effluent that was supposed to be in the vessel. It would have disposed tidily of an awkward situation, had it not been accidentally discovered by the Greenpeace Geiger counters. To be sure the discharge was contrary to the terms of BNFL's site licence; indeed the whole exercise had such an unpleasant taste that it was referred to the Director of Public Prosecutions. After due deliberation he announced that criminal charges were to be laid against BNFL. At the time of writing the case had yet to come to trial.

Meanwhile, the charges that BNFL was laying on its customers were arousing aggrieved recriminations. In December 1983 it was reported that BNFL had unilaterally announced a 30 per cent increase in the cost of reprocessing, to clients who had already signed contracts at the earlier price. The contract terms gave BNFL the right to impose the increase; but the clients were less than happy. One that had signed a contract only a few months before was quoted as saying that if it had known of the impending increase it would not have signed the contract.

As for THORP, in which the contracted reprocessing services were to be carried out, BNFL did not even apply for detailed planning permission for the plant until early 1983. The local council then temporarily withheld planning permission: BNFL had still to satisfy the council's demand for improved road and other related facilities. BNFL was also refused permission to extract process water from the most spectacular of the lakes in the adjacent Lake District, the magnificent Wastwater. By this time the new oxide reprocessing plant was three years behind the schedule BNFL had put forward at the Windscale inquiry. Despite BNFL's ebullient confidence, few informed observers were betting on when THORP would actually come into service, how it would work, or indeed if.

PART III
Fast Breeders in Neverland

9 Slow starters

The free range of discussion at Harwell was symbolized by the establishment of a 'Crazy' Committee, which met only twice and achieved nothing. There was also a Harwell Power Committee, for discussion with outside scientists and industrial engineers, but this soon petered out and was replaced by Harwell Power Conferences (held in 1948, 1949, 1950, 1951 and 1953) which met for a day or two to inform, and consult with, a few selected outsiders mostly from industry. Harwell's main forum for discussion was its own Power Steering Committee, which was set up to examine every known aspect of power generation from first principles and to plan a rational research and development programme. It was to be the sole authorizing agency for experimental work on power and the clearing house for all new ideas.

So says Margaret Gowing's masterly official history of postwar nuclear activities in Britain, *Independence and Deterrence*. The history then notes drily that:

for two or three years the Harwell power debates ranged over a much wider front, shifting first in one direction, then in another. The minutes of the Power Steering Committee referred frequently to 'prolonged', or 'heated', or 'inconclusive' discussions and Cockcroft [Sir John, Harwell director] had difficulty pulling a clearly defined programme out of the debates.

Does this not sound curiously familiar? It is clear that Britain's interminable wrangle over reactor choice had its foundation – if that is the correct term for the quagmire that ensued – in the 1940s. Gowing then adds a revealing comment: 'The only point on which there was general agreement throughout all these years was on the long-term future – on the ultimate and overriding importance of breeder reactors, which would produce more secondary fuel than the primary fuel they consumed.'

The reason for this island of unanimity amid the prevailing conflict of views was straightforward. In the late 1940s and early 1950s uranium was scarce and expensive; moreover its supply was politically acutely sensitive, because of the weapons implications. Cockcroft spelled out the consequences in a lecture entitled 'The Development and Future of Nuclear Energy', delivered on 2 June 1950. As he saw it, the long-term objective was

> to build nuclear power stations which will produce power at a cost not very different from a coal-fired station. For this to be worthwhile we must have adequate uranium-ore reserves in sight to fuel our nuclear power stations for many centuries . . . For this we have to develop a new type of atomic pile [reactor] known as the 'breeder pile' because it breeds secondary fuel [plutonium] as fast or faster than it burns the primary fuel uranium-235 . . . These piles present difficult technical problems, and may take a considerable time to develop into reliable power units. Their operation also involves difficult chemical engineering operations in the separation of the secondary fuel from the primary fuel.

Cockcroft's reservations were all too justifiable. Part II of this book has already described subsequent British experience with the chemical engineering – that is, reprocessing – in question. The following pages will discuss the 'considerable time' taken by Cockcroft's 'breeder piles' to 'develop into reliable power units'. The time so far is thirty-five years and counting.

To be sure, Cockcroft was not the only one with some early reservations about the real prospects for breeder reactors. Harwell was home to the nuclear physicists, whose interests had a strong tendency towards the theoretical. Risley, under the leadership of Christopher (later Lord) Hinton, was home to the nuclear engineers, whose perspectives at the time focused more on nuts and bolts, and how to attach the one to the other for a practical purpose. The Risley engineers did not share Cockcroft's conviction about the breeder as the key to future nuclear power, as they made clear in 1953. By this time they had been working for some two years with their Harwell colleagues on the design of a full-scale fast breeder power station. The Risley engineers summed up their feelings succinctly in a report on their efforts: 'At first sight this fast reactor scheme appears unrealistic. On closer examination it appears fantastic. It might well be argued that it could never become a serious engineering proposition.'

Nevertheless, in only two years it had, in official eyes at least: construction work started in March 1955 on an experimental fast breeder power station at the new Atomic Energy Authority's new Dounreay Experimental Reactor Establishment, on the north coast of Scotland not far from John o'Groats. This remote location was chosen precisely for its remoteness. There were as yet major questions unanswered about the possible range of behaviour – and misbehaviour – of a reactor whose core contained an unprecedented concentration of fissile material. In the words of the first annual report of the AEA, in 1956:

> Owing to the small core size, there is a risk that an accident in the reactor might lead to a rapid rise in temperature which in turn might cause the melting of the fuel elements. If this should happen there might be an escape of fission products from the core. To prevent these from being dispersed outside the reactor, it will be enclosed in a steel sphere about 140 feet in diameter.

The spherical dome of the Dounreay Fast Reactor (DFR for short) was to become one of the best-known images of British nuclear power.

By the time of the third annual report, in mid-1957, the DFR was 'expected to start operation in 1958'. From 8 October 1957, however, the AEA was more than somewhat preoccupied with the Windscale fire and its aftermath*. Work at Dounreay suffered from this 'diversion of effort', as the 1958 annual report put it. Accordingly, 'the date when the fast reactor will become critical has been postponed from 1 April 1958, for several months'. Given the usual associations with 1 April it would not in any case have been a propitious choice of date for start-up. In the event the reactor did not actually go critical until November 1959. The 1960 annual report, recording this landmark, went on to remark that 'A prototype power-producing reactor may be built for operation about the year 1967, the development of which will enable a commercial power station to be specified.' The proposed prototype was indeed built – but not for operation in 1967.

The design output of the DFR was intended to be 60 megawatts of heat, or in due course 14 megawatts of electricity. Successive AEA

* See *Nuclear Power*, Walter C. Patterson, Pelican, 1983.

annual reports stressed that the DFR was 'experimental', 'intended to develop the technology of fast reactors generally'. It fulfilled this role admirably, in that it succumbed to a fascinating variety of novel engineering difficulties, particularly those arising from the use of molten sodium metal as the cooling fluid. By mid-1961 its highest output had been 1.5 megawatts of heat. Her Majesty the Queen Mother visited Dounreay on 14 August 1961, to inaugurate the resumption of experimental work on the DFR after extensive modifications. By mid-December the reactor had been run up to 11 megawatts of heat, at which point it was shut down to have its fuel core replaced with one of improved design. While thus busy with the DFR the AEA in 1961–2 was also completing a reference design study for a 500-megawatt fast breeder power station. The next step in the programme would be to try out the concepts on an intermediate scale, on what would be known as the Prototype Fast Reactor.

The Dounreay Fast Reactor reached an output of 30 megawatts of heat – half its intended design output – on 7 August 1962, and remained at this level for the rest of the year; in October it supplied electricity to the national grid for the first time, albeit strictly on a by-product basis. New fuel designs were under continual development and testing, aimed at higher output and also at higher 'burnup' per fuel element.

The Select Committee on Nationalized Industries, in its May 1963 report on the electricity supply industry, noted that 'the development by the [Atomic Energy] Authority of a fast breeder reactor at Dounreay . . . remains a long-term project. The Authority hope that a prototype will be operating by 1969 or 1970; and the first civil station would not be working before 1975'. Be that as it might, the 1963–4 annual report of the AEA declared that 'Consortia design engineers are engaged on a design study of a 1000-MW(E) [megawatts electric] power producing fast reactor'. At the time the largest thermal reactors contemplated for construction were of 660 megawatts electric. In the meantime, in July 1963, the Dounreay Fast Reactor at last attained its full design output of 60 megawatts of heat, or 14 megawatts of electricity, and operated at this level for most of the ensuing year, interrupted only by shut-downs to refuel or to examine experimental fuel. From that time on the DFR, in its capacity as an irradiation and fuel-performance research facility, served well for several years, while the AEA pressed on with work on its successor, the Prototype Fast Reactor.

In the course of 1964–5, while much of the AEA's attention was focused on the battle between light-water reactors and advanced gas-cooled reactors for the second nuclear power programme, the AEA nevertheless completed not one but two design studies for fast breeders. The first was for the proposed Prototype Fast Reactor or PFR. It was to have an output of 600 megawatts of heat or 250 megawatts of electricity; but it was designed to use components suitable for a full-scale commercial fast breeder power station. By the time of the AEA's eleventh annual report, in August 1965, AEA staff were already preparing detailed designs and specifications for major plant and civil engineering contracts for PFR. This was, to be sure, well before the official go-ahead for PFR; it did not come until 9 February 1966, in a statement to the House by Minister of Technology Frank Cousins. PFR was to be sited at Dounreay, next to the operating DFR.

The announcement disappointed at least one group. Within the AEA there was a strong faction eager to build the new PFR at the Authority site at Winfrith Heath, home of the Dragon high-temperature reactor and the steam-generating heavy-water reactor. The AEA wanted to demonstrate that a fast breeder need no longer be exiled to the most remote tip of the British Isles for safety reasons. However, the Labour government's decision to put PFR next door to DFR had less to do with the safety of the populace than with the safety of the Parliamentary seat of Caithness and Sutherland, of which Dounreay was a conspicuous geographical and industrial landmark. In short order the locals demonstrated that a fast breeder power station introduced by a Labour government could be hazardous to a Liberal MP. At the general election less than two months after the Cousins announcement the seat went to Labour.

There was a measure of rationality about the choice of Dounreay as the site for PFR. The site already had extensive experience of a fast breeder; moreover it also had a pilot-scale reprocessing plant designed to accept the distinctive fuel used in a fast breeder, with its high concentration of fissile material. Once separated, nevertheless, the recovered plutonium would have to be returned to Windscale for fabrication into fresh fuel, to the plant that would manufacture all the fuel for the reactor. There was also one other question, of a more fundamental nature, about the site. Dounreay was some seventy miles north of Inverness, the nearest city of any size in the region – and accordingly the nearest major load centre that could use the 250 megawatts of electricity to be generated by PFR. Even getting the electricity from PFR to Inverness would be a substantial and expensive undertaking; it would also entail significant and costly transmission losses. Given the essentially experimental nature of PFR the question was not at the time pressing. In due course, however, in regard to PFR's proposed follow-up, the question was to gain much more prominence: does it make sense to site a 1200-megawatt power station more than seventy miles from the nearest load centre – even to win a seat in Parliament?

10 The ten per cent reactor

The inner and outer breeder sections of the DFR were originally loaded in 1958 with natural uranium elements clad in stainless steel. Early in 1965 it was found that a few of the lightly irradiated elements in the outer breeder were difficult to remove, although the inner breeder elements were in good condition. A comprehensive survey of the outer breeder was carried out in September 1965, and a number of elements were found to be distorted or swollen. Investigation showed that this had been caused by higher than normal uranium temperatures due to abnormal coolant flow conditions in some regions of the breeder. This will not occur in future fast reactors since coolant flow conditions will be different, and the breeder fuel itself will be ceramic and therefore not subject to the temperature limitations of natural uranium. It was decided to remove 500 breeder elements, and to carry out the work special cutting tools and removal equipment had to be manufactured. The work was completed by the end of December and the reactor went critical again on 23 January after loading new experiments. (AEA Annual Report 1965–6, paragraphs 139–40)

AEA fast breeder people took technical hiccups like this very much in their stride. That was what the DFR was for. Such incidentals in no way weakened their confidence in the concept of the fast breeder. On the contrary: while they pressed on with detailed designs for the Prototype Fast Reactor they had already satisfied themselves that the prospects were excellent:

The design study of a 2 × 1000 MW(E) fast reactor power station in general endorsed the conceptual design of the prototype fast reactor as representing the most likely features of the first commercial fast reactors. A capital cost estimate for this study indicates a cost similar to that of the best thermal reactor available at the same time, with potential for further reductions. (AEA Annual Report 1965–6, paragraph 157)

The Select Committee on Science and Technology was duly impressed. In its first report, in October 1967, it noted that

The fast 'breeder' reactor is the system on which the long-term prospects of nuclear energy generation are based ... Work on this system has been increasing steadily for some ten years and the greatest effort of the AEA's research and development programme is now devoted to this type of reactor. Expenditure in 1966–67 was about £12 million and there will be increasing capital expenditure over the next few years as the construction of the Dounreay fast reactor prototype proceeds. This system is regarded as likely to provide a very cheap source of electricity. Building costs (at 1967 prices) of fast reactor stations are expected to be as low as £50 per kilowatt installed and generating costs to be reduced ultimately to 0.3d [old pence] per kilowatt hour. The prototype – a large station producing 250 MW(E) – is expected to be on power in 1971.

This expectation proved to be more than somewhat sanguine.

Nevertheless, by 1968, with the PFR still not much more than a hole in the ground, the AEA was looking to have at least 15,000 megawatts of fast breeders in operation by 1986. On the basis of 'another bold decision' by government, exploitation of the fast breeder would be 'the major event of the rest of the century'. By 1969 the AEA was asserting that: 'The UK has the firm intention of introducing fast reactors as rapidly as possible after the operation of our 250-MW prototype.'

Meanwhile, back at Dounreay, all was not quite as well as these confident pronouncements suggested. In May 1967 the primary cooling circuit of the DFR sprang a leak. Although the AEA described the leak of molten sodium as 'small', the reactor was shut down in July 1967 for repairs that kept it out of action until late June the following year. Work on the construction of the PFR also ran into trouble. The complex rotating 'roof' of the reactor, from which were suspended important components of the reactor internals, was proving much more difficult to fabricate successfully than the AEA had anticipated. In consequence the AEA conceded that the PFR would not be in service in 1971 as earlier claimed. The AEA was not apparently much bothered by this, declaring with blithe insouciance that:

Trouble was experienced in welding the biological shield roof and it has not yet been delivered to Dounreay; construction of PFR has been delayed by about a year. The roof is of conventional engineering structure and the troubles which have been encountered are not connected with fast reactor aspects. Their recurrence in future reactors can be avoided by detailed changes in design and manufacture.

The AEA comment prompts an obvious question: if the roof was just a 'conventional engineering structure', why didn't they get it right the first time? In due course the PFR was to give the AEA ample opportunity for such airy off-hand alibis.

The DFR, meanwhile, was continuing to fulfil its experimental role, by manifesting further categories of engineering problems for fast breeders. One in particular proved knotty. No reactor hitherto in operation had subjected its structural materials to intense high-energy neutron radiation for lengthy periods. It had been anticipated that the extremes of irradiation would make fuel pellets swell and distort, as a result of fission and neutron bombardment. In fact the fuel material itself proved to be less of a problem than the structural materials containing and supporting it. The fast neutrons knocked atoms out of place in the stainless steel, leaving 'voids' that weakened the crystal structure and deformed the components made from it. The problem was not really a surprise, or unexpected; but it posed a daunting challenge to the metallurgists and the design engineers. They had to find materials able to withstand the demanding environment in the core of a fast reactor, and arrange them with adequate tolerances to accommodate the subsequent changes of shape and strength. The void-formation problem loomed large in the context of design engineering for the PFR. It was one factor that prompted the decision to settle for a less strenuous operating regime in the PFR than in the DFR, with lower temperatures and power densities.

While these practical problems occupied the attention of the staff at Dounreay, the AEA was linking up with the CEGB and the two reactor-building consortia for further design studies on commercial fast breeder power stations. On 14 October 1970, introducing the AEA annual report, chairman Sir John Hill characterized the outcome thus:

On the planning side of our fast reactor programme we have had a most useful study of the fast reactor by a group consisting of engineers of the CEGB, the industrial design and construction firms and the Authority. As a result we have now an agreed programme which is being undertaken by all parties which could lead to the CEGB being able to start the construction of the first civil fast reactor, possibly of 1300 MW, by early 1974. This is, of course, in no way a commitment to proceed on this timescale, but rather a basis of programme planning which will be subject to regular reviews of progress and, of course, seeing how the prototype fast reactor performs in 1972 and 1973.

As it turned out, the prototype fast reactor did not perform at all in 1972 and 1973. Nevertheless, a year after Hill's comments quoted in the preceding paragraph, and with the PFR falling steadily farther behind schedule, the 1971 AEA annual report was still confident. The cooperative study aforementioned had resulted in

the formulation of a strategic plan for the introduction of fast reactors to the CEGB network; this assumes that construction of a first commercial station will start in 1974 as a 'lead' station, following operation of the PFR. This would be followed by other stations after an interval of perhaps two years. This plan assumes that the technical and economic results from the development programme confirm present expectations; it will be reviewed each year in the light of progress achieved.

If such reviews in due course took place they were remarkably oblivious to the widening divergence between the AEA's expectations and technical and economic reality.

In October 1971 the AEA's 'present expectations' were to say the least robust:

It is estimated that, in only thirty years from now, over three-quarters of all electricity in the United Kingdom will be generated from nuclear power and that more than half of this nuclear generation will stem from fast breeder reactors (to the development of which almost half the effort on the Authority's reactor programme is currently geared).

Sir John Hill had already expounded to his fellows at the fourth UN Conference on the Peaceful Uses of Atomic Energy in Geneva in September 1971 the 'strategic plan' endorsed by the electricity authorities, the nuclear power industry and the AEA.

By 1979–80 we should have had seven years' operating experience with the PFR, constructional experience of perhaps three or four large commercial stations, and initial generating experience from the first of these larger units. On this basis we would expect that by about 1980 we would have sufficient confidence and experience to incorporate fast reactors into the United

Kingdom generating system to the maximum extent consistent with the availability of fissile material and the growth of demand for new generating plant. Whether such a timetable can, in fact, be achieved will depend on technical developments over the next few years. This, however, is the plan to which we are working and so far we see no reason why it should not be achieved.

Reasons there were to be, in abundance; but the AEA refused to see them, even when it tripped on them and fell flat on its face.

These confident pronouncements were being uttered against the background of nuclear power policy described in Part I. Dungeness B was in chaos; the later AGR stations were already falling behind schedule. The consortia had dwindled to two; the choice of reactor for forthcoming nuclear stations was under examination in the secret and eventually fruitless deliberations of the Vinter committee. The Vinter committee examination of thermal reactor policy was paralleled by another on fast reactor policy. Its findings, like those of the Vinter committee, were never to be made public; but they undoubtedly underpinned the relevant part of the policy statement delivered by John Davies, Secretary of State for Trade and Industry, on 8 August 1972. Davies told the House of Commons that the government wanted 'to push ahead as rapidly as possible with development of the fast reactor'.

'As rapidly as possible' was not, however, as rapidly as expected. Sir John Hill conceded this in his remarks on 5 October 1972, introducing the Annual Report. About the still-unfinished PFR he said:

We expect the reactor to be producing electricity by the end of 1973. We in the Authority have never proposed that the first commercial fast reactor should be started until sufficient operating experience of the prototype had been obtained, to be absolutely sure that there were no fundamental problems unresolved. I have, however, always believed in continuity of design and experience and would like to see the next reactor started as soon as the lessons of the first have been fully assimilated by the designers and engineers. Clearly our hopes of a 1974 start are now too optimistic in the light of the commissioning and operating dates for the prototype and the amount of component testing now judged necessary. The design of the CFR is, however, under way . . .

'CFR' was the latest acronym, standing for Commercial Fast Reactor. It was soon to undergo a subtle but revealing change.

The February 1973 issue of *Atom*, the AEA's popular monthly, reported on a meeting the previous November, attended by senior civil servants and nuclear industry management, on 'Future Prospects for Energy Supply and Demand', presented by the 'New Systems Forum' of the AEA. According to the report of the meeting: 'A commercial fast breeder power station programme commencing with a lead station coming on line in 1981 and further stations in the mid-1980s appears to be a reasonable assumption on the basis that PFR know-how and experience will be adequate for a first order to be placed for around 1976.' The almost imperceptible note of caution – 1976, not 1974, and the 'mid-1980s' for subsequent stations – had to be set against the assumption that a station ordered in 1976 could be 'on line in 1981'. This allowed only five years for construction and commissioning, compared to the eight-plus years already run up at Dungeness B and at contemporary fossil-fuelled stations likewise still unfinished.

Even the faint note of caution in this report was swept aside in an aggressive presentation delivered in the US in mid-1973 by Tom Marsham, deputy managing director of the AEA's Reactor Group:

> Satisfactory experience with the experimental reactor DFR in the early 1960s led to construction of the 250 MWE power station at Dounreay which will be brought to power this year. Some two or three years from then, we are expecting to start constructing the 1300 MWE lead commercial station with ordering of subsequent commercial plants building up to large scale during the early 1980s. There is nothing adventurous or foolhardy about this plan.

Nothing, perhaps, except its central premise. The end of 1973 arrived and departed with the Prototype Fast Reactor still awaiting its first criticality, to say nothing of being 'brought to power'. One primary and one secondary sodium pump malfunctioned during tests; both had to be removed from the reactor for detailed examination. Tests continued with the remaining two primary pumps in place. On 11–14 March 1974, however, the British Nuclear Energy Society was to play host to a major international conference on 'Fast Reactor Power Stations', with delegates from France, the US, the rest of Europe, Third World countries and even the Soviet Union. The ignominy of welcoming the foreign visitors to the conference with the PFR still cold was too much to contemplate. The week before the conference the AEA pulled out the control rods at Dounreay, and on 3 March 1974 started up their new reactor for the first time. On the opening day of the conference they announced the fact with pride; it was thus far from coincidental that their French colleagues duly announced, on the closing day of the conference, that the French Phénix fast breeder had just attained full power. Nuclear one-upmanship has many facets.

The conference heard many papers from fast breeder engineers, extolling their successes and discounting their difficulties. One paper in particular, however, attracted a measure of attention in the context of the British fast breeder programme. It was delivered by Eric Carpenter, head of reactor physics at the CEGB's Berkeley Nuclear Laboratories; and it warned that the CEGB was less enthusiastic than the AEA about a rapid move into fast breeders. Reliability was a crucial factor; together with delays in construction, lack of reliability had 'a much bigger deleterious influence on electricity costs than almost any of the advantages claimed in the brochure assessments'. The CEGB by this time had all too much first-hand experience of both delays and unreliability of its conventional nuclear stations, and of what the paper scornfully called 'brochure assessments' – a slightly less dismissive label than 'back-of-the-envelope' but clearly carrying the same pejorative import. The paper asserted that the putative savings accrued by introducing fast breeders as fast as possible would be no more than 5 per cent of total expenditure on a nuclear system – and then only in what it called 'the unlikely event of capital costs of fast and thermal reactors being equal'. The CEGB contributors considered that no order for a fast breeder power station could be placed before 1977 or 1978 at the very earliest. The CEGB's concern about failure to meet construction schedules for fast reactors was already underlined by the PFR's record to date: expected on line in 1971, it had not even gone critical until March 1974. Worse, however, was to come.

Throughout the summer and autumn of 1974 staff at Dounreay continued running the PFR at low power. Small leaks appeared in the steam-generators, the distinctive 'boilers' in which hot molten sodium passed through thousands of fine tubes to boil the water around the tubes. Such leaks were a particular problem in a sodium-cooled system, because of sodium's well-known eagerness to react chemically with water. A major leak, like one that had happened at the Soviet fast breeder prototype at Shevchenko in November 1973, would release enough hydrogen and heat to create a serious hazard of explosion; even a minute leak, invisible to the naked eye, would lead to the formation of hydrogen bubbles in the sodium coolant, presenting at the very least an unwelcome irregularity in the cooling-flow, and possibly actual control problems. Accordingly, the leaks in the PFR steam-generators gave rise to concern. By the end of October 1974 the most troublesome unit was decoupled from the reactor in order to seek out the leaking tube to plug it.

The rest of the reactor seemed to be behaving well; and the AEA therefore laid on a major press visit, flying some seventy journalists to Dounreay on 30 October. Journalists found the outing enjoyable and instructive, but wondered why the AEA had chosen that particular time to invite them, since nothing of any special moment occurred during the visit to the site. It transpired subsequently that the AEA had apparently been expecting to switch power from the PFR into the national grid. In the preceding week, however, there had been a fierce storm in the North Atlantic. The storm had uprooted hundreds of tonnes of seaweed, which had been sucked into the reactor's cooling-water intake just off-shore. The generating set had had to be shut down while Dounreay staff with long-handled rakes cleared the obstruction from the intake. Since this was not quite the story the AEA hoped to put over, nothing was said about it to the visiting journalists.

Six months later the PFR once again played host to a visiting party. At the end of April the newly-formed European Nuclear Society held its inaugural conference in Paris; after the conference one of the side-trips laid on took nuclear people from all over Europe to Dounreay. AEA staff were happy to show off their reactor, which was, they said, working fine; a month earlier the plant had generated its first electricity. Unfortunately, however, it had yet to reach a power level above 12 per cent of its full thermal capacity. Small but persistent leaks in the sodium-water steam-generators kept two of the reactor's three cooling circuits out of operation. PFR staff carried on operating the reactor on its one remaining cooling circuit, but trouble with turbine bearings interrupted even this limited operation. Then, just before the nuclear dignitaries arrived from Paris, more small leaks manifested themselves, this time in a section of the only operative cooling circuit.

The AEA staff at Dounreay put on brave faces, but the ENS visit cannot have been an especially happy occasion. As one sympathetic visitor, the editor of *Nuclear Engineering International*, put it: 'Thus, although the reactor itself has been operating very well it has not yet been possible to build up any significant amount of fuel irradiation.' Nor, it might be added, to generate any significant amount of electricity, a point that probably had not escaped the notice of the Central Electricity Generating Board. The AEA was to continue to protest that the reactor itself was working well, and that the stubborn troubles at Dounreay were with the generating set and the steam-generators. This was undoubtedly true, for what it mattered. But the CEGB had already suffered many years of frustration with its own generating sets, and knew what a headache these could be.

Furthermore, to suggest, as the AEA was trying to suggest, that the steam-generators were somehow ancillary, not part of the nuclear system, was indefensible special pleading. One of the unique distinguishing characteristics of the fast breeder design selected by the AEA was precisely the choice of molten sodium as a coolant. If you could not then use the molten sodium reliably to boil water, you had a basic design problem – one that could not be brushed aside by reference to the 'satisfactory' operation of the reactor core itself.

As 1975 ticked away the AEA voiced hope that the PFR would at last reach full power early the following year. However, by February 1976 *Nuclear Engineering International* was dashing cold sodium on the possibility:

> Hope that the Dounreay Prototype Fast Reactor (PFR) would be brought up to full power in February will not now be fulfilled. The designed output of 250 MW(E) is not now likely to be achieved 'for several months'. The reactor continues to operate satisfactorily and with number 1 secondary (cooling) circuit in operation an electrical output of 40 MW(E) has been achieved with a thermal power of about 200 MW (of heat) . . . Work in preparation for recommissioning of number 3 secondary circuit is well advanced. The circuit has been filled with sodium and clean-up operations are in progress . . . It was expected that this circuit would be available for power operation during the next few weeks. On number 2 circuit, work on checking the superheater and to determine how best to operate has progressed well.

As well, perhaps, as could be claimed, given that all of this was supposed to be known before the reactor was built. The following month *Nuclear Engineering International* offered an oblique compliment of sorts; in its annual world review of the status of reactors still under construction or being commissioned, it no longer included the PFR. In the light of subsequent developments at Dounreay, *Nuclear Engineering International* spoke – or rather remained silent – too soon.

By September 1976 some of the news from Dounreay, as noted in *Nuclear Engineering International*, was at last genuinely good: 'During most of August the 250 MW(E) prototype (fast breeder reactor) at Dounreay has been operating on all three of its coolant loops with all of the early heat exchanger problems now remedied. The maximum power reached so far is 500 MW (of heat), but full power was expected to be reached by the first week in September.' However, the report continued with additional news of a slightly more disconcerting kind, albeit presented as a straightforward matter of fact: 'Plans to replace all three types of heat exchanger with improved designs using austenitic steel and avoiding the thick tube plates where corrosion has occurred are still proceeding as scheduled for installation in 1979.' When this 'schedule' for replacing major plant components with completely new ones had been decided the magazine did not say. It was nevertheless a further indication that the PFR was a lamentably long way from demonstrating that fast breeders could fulfil the CEGB's requirements that they be reliable, built on schedule and within budget.

The AEA remained cheerfully confident about the PFR. A measure of its confidence was the decision to shut down permanently the little DFR next door. On 23 March 1977 Lord Hinton, who had chosen the Dounreay site and supervised the early stages of construction of the DFR, threw the switch that consigned it to history. His reflective remarks on the occasion, reprinted in the AEA monthly *Atom*, were a *tour de force* of personal reminiscence interspersed with incisive views on the current state of the art, including the Prototype Fast Reactor.

I hope and believe that many lessons have been learned from PFR. At one of the early Fast Reactor Design Committee meetings Jim Kendal, whose feet were usually very firmly on the ground, put forward a complicated proposal for the design of the fast reactor and I remember saying to him, 'Look Jim, that's a very clever idea but I don't pay you to be clever, I pay you to be successful'. Most of the mistakes (and fortunately they have been rectifiable) on PFR have been made because engineers have thought they were just that little bit more clever than any of us really are.

It was by Hinton's standards gentle chastisement; Hinton went on to give firm endorsement to the proposal to build a full-scale fast breeder 'not later than the end of this year ... the aim should be to commission it before 1985'. Unfortunately, Hinton's assumption about the ready rectification of the mistakes on the PFR was entirely too premature.

As the months and years passed the AEA's defiantly laudatory animadversions to the PFR, in lectures, papers and annual reports, sounded rather like the *Punch* curate's acclaim for his egg: 'Parts of it are excellent.' When the AEA first revealed its intention to replace most of the PFR's heat-exchangers, reports said that the replacements would be in service by 1979. They were not. Over the years, periodic questions in Parliament elicited monotonously similar answers: the cumulative 'capacity factor' (output of electricity from the PFR as a fraction of its design capacity) remained stuck year after year at about 10 per cent. In October 1984 the authoritative quarterly analysis published in *Nuclear Engineering International* gave the total lifetime capacity factor of the Prototype Fast Reactor in the first ten years after its start-up as 9.9 per cent. As a response to the CEGB's call in February 1974 for reliability and adherence to schedules, the PFR was looking like the fast-neutron answer to Dungeness B.

11 Fast or bust

In the late 1960s the Atomic Energy Authority was nurturing five different types of reactor: Magnox, AGR, high-temperature, steam-generating heavy water, and fast breeder. A decade later the number had dwindled to one. The Magnox reactors had reached the end of the line. Except for occasional trouble-shooting like that on pipework welds, the AEA had little further interest in Magnox reactors. The AGRs were still keeping their heads above water; but the main responsibility for further design engineering now lay with the National Nuclear Corporation. The high-temperature reactor was dead; the steam-generating heavy water reactor was doomed to extinction as the first and last of its line. Of all the AEA's nuclear progeny only the fast breeder survived under its wing. Needless to say the AEA fussed over the fast breeder like a mother hen over its last chick. But the chick was looking more and more like an ugly duckling. The AEA remained convinced that it would eventually prove to be a swan. Others were beginning to suspect that it was actually a turkey.

This heretical thought had taken a long time to surface. In the mid-1960s official opinion, led by the AEA, assumed without question that a rapid progression from the little Dounreay Fast Reactor to the larger Prototype Fast Reactor to a series of full-scale fast breeder power stations was not only natural but obviously desirable. The only possible constraint foreseen was a conceivable shortage of plutonium to fuel the full-scale fast breeders; with that in mind it was the reiterated policy of government and AEA to reserve all 'civil' plutonium separated from CEGB and SSEB spent fuel, against its imminent use to fuel the coming fast breeder power stations.

The PFR had been expected to be on line by 1971, paving the way for an immediate start on its successor, known as the Commercial Fast Reactor or CFR. By late 1971 it was abundantly evident that the

PFR would not even be completed, much less on line, for many months to come. As described in earlier chapters, AEA chairman Sir John Hill was not, however, troubled by this. On 8 August 1972 John Davies, Secretary of State for Trade and Industry, in his statement to the Commons on nuclear power policy, once again gave the ritual blessing to the fast breeder. The AEA's other reactors might be coming in for ever more sceptical scrutiny, but the fast breeder was sacrosanct. Even in 1975, when the PFR had at long last gone critical only to manifest the leaks that would cripple it, the official commitment remained unshaken.

Just how committed could be seen from the AEA's evidence to the Royal Commission on Environmental Pollution – the so-called Flowers commission described earlier. In September 1975 the AEA submitted a paper to the Flowers commission taking as its premise a nuclear programme that would have a total of 104,000 megawatts of nuclear power in operation by the year 2000, of which no less than 33,000 megawatts would be fast breeders. At the time the total operative nuclear generating capacity in Britain was less than 5000 megawatts; the nuclear plant construction industry was in chaos; and the PFR had yet to attain more than a modest fraction of its intended design output. Sir Brian Flowers, himself a part-time Member – board member – of the AEA, was reported to have taken exception to this scenario as being utterly unreal.

The AEA hastily insisted that it was not a forecast, merely a 'reference programme' to establish an upper limit on the scale of British nuclear involvement for purposes of weighing environmental impact. Be that as it might, the AEA clearly considered this 'reference programme' as plausibly achievable. As late as January 1977, at a public one-day conference on the fast breeder at Imperial College under the aegis of its Rector, Sir Brian Flowers, one of the AEA authors of the 1975 paper doggedly defended it. In his view the programme envisaged could be readily achieved, simply by relying on the potency of exponential growth – one plant, then two, then four, until they were springing up like mushrooms. Other conference delegates, mindful of the actualities that overrode such abstractions, remained to say the least dubious.

The September 1975 AEA evidence was by no means the last hearty official endorsement of the central role of the fast breeder in

Britain's energy planning. In early 1976 the Advisory Committee on Research and Development for Fuel and Power, known as ACORD, demonstrated that it was indeed in full accord on the fast breeder. The committee was drawn from the senior executives of management and labour in the fuel and electricity supply industries, and chaired by the Chief Scientist in the Department of Energy. In early 1976 the Chief Scientist in question was Walter Marshall, who also happened to be deputy chairman of the AEA. Under Marshall's chairmanship ACORD drafted a report on 'Energy Research and Development in the United Kingdom'. The ACORD report assigned priorities to different categories of energy R&D, from one star to five in order of importance. To no one's surprise ACORD accorded the fast breeder five stars across the board.

Since the beginning of the 1970s the AEA had been pleading for government permission to build its long-awaited Commercial Fast Reactor. Design teams from the AEA, the CEGB and the nuclear plant manufacturers had been busying themselves for years laying out their paper power plant, based on a 1200-megawatt fast breeder. By 1976 the AEA was spending close to £100 million a year on the fast breeder – not in major capital investment, just in funding this accelerating research and development. In 1976, especially after publication of the ACORD report, confident rumour had it that the go-ahead for the CFR was at last imminent.

The rumour had received a boost from the suggestion that the Flowers commission would be advocating CFR. At the end of 1975, however, Sir Brian Flowers declared that this suggestion was 'quite false'. Flowers published letters he had exchanged with Prime Minister James Callaghan, asking that the government hold off any decision 'on whether to proceed with such a plant in collaboration with other European countries' until the commission had published its report some months later. Failing such a postponement the commission wanted to see a clear distinction drawn between a single full-scale demonstration fast breeder and a large continuing programme of such plants. The commission conceded that by building one full-scale plant Britain might contribute significantly to resolving what the commission called 'the serious fundamental difficulties' associated with the fast breeder. However, no official body had for many years so much as hinted that the fast breeder could even raise 'serious fundamental difficulties'. What these difficulties might be Flowers indicated indirectly in his letter:

> The demonstration site should be remotely sited; it should have its own fuel reprocessing and fabrication plant on site in order to remove the security risks of shipment of plutonium; it should be provided with every means of protection, including both physical devices and an armed security force; and experience of plutonium accountability and inspection should be designed into its system.

It was not exactly a reassuring recipe.

On 22 June 1976, at Energy Secretary Tony Benn's National Energy Conference, Flowers was more specific about the commission's unease about the use of plutonium as a civil fuel, as described in Part II above. Earlier in June Benn had told the Commons that the government would announce in the early autumn its decision about the future of Britain's fast breeder programme. Work had reached a point at which the government had to decide

> our approach to the next stage of the system's development, including our policy on the construction of a full-scale demonstration reactor. This is a matter of great public importance in terms of long-term energy provision and the safety and environmental considerations. In my current review of this I wish to provide the opportunity for wide consultation. I shall take full account of the prospects for international cooperation and the forthcoming report on radiological safety from the Royal Commission on Environmental Pollution.

As it turned out, however, the Royal Commission was concerned about more than just radiological safety. As described earlier the commission was deeply apprehensive about the implications of a commitment to what it called the 'plutonium economy'. It accepted that there was a case to be made for building a single large fast breeder, to assess its safety and social implications. But the commission went on to warn that 'we must view this highly significant first step with misgivings ... The strategy that we should prefer to see adopted, purely on environmental grounds, is to delay the development of CFR1' (paragraphs 517–18). From the day of publication of the Flowers report (22 September 1976) onwards, the prospect for even a single large fast breeder in Britain looked distinctly bleaker.

In a television interview not long afterwards, Energy Secretary

Tony Benn acknowledged that the government had promised a decision on the fast breeder by the autumn; but he now felt more time was needed. The view that there was no alternative to a fast breeder programme was, he said, an argument that had to be looked at very carefully indeed. Neither he nor the government accepted that a fast breeder programme was absolutely inevitable. No single expert view could answer all the questions. If ever there was an area, he said, in which long-term planning was essential, it was in energy. He could think of no decision of a government that so committed a nation over thousands of years.

This perspective of millennia may have been one of the reasons that prompted the British Council of Churches to hold two days of hearings, 13–14 December 1976, into the social, political and ethical dimensions of fast breeders. It was to say the least an unusual intervention by the clergy; but both the organizers and their invited witnesses took the hearings very seriously. One witness was Benn himself; and he made his own position clear:

> There is another set of factors to which reference has been made in public debate: I would describe them as domestic political factors arising out of two considerations. One is the problem of security and the risk of terrorism and the second arises from what happens when you have policies so complex that the democratic process finds it hard to come to terms with the choices that have to be made. Certainly as a Minister with these responsibilities now on and off since 1966 when I first became Minister of Technology, I have always found nuclear policy the most difficult because Ministers are not experts, they are not scientists, they are not engineers, they are not qualified to assess in any way the technical decisions that had to be made. And yet, whether you look at it in terms of the environment or safety or energy policy, or the massive public expenditure involved in all the projects of this scale, it is essential that nuclear policy be preserved within the democratic framework of control and not subcontracted off to those whose only claim to reaching decisions might rest upon their technical qualifications. I think it would be very frightening indeed if we were to say that our fuel policy required us to adopt a technique of production like nuclear power which in its turn required the decisions to be taken from the process of government answerable to Parliament and the public, and put into the hands of those whose special qualifications for deciding them would rest upon their technical knowledge.

In the weeks that followed, moves by US President Jimmy Carter, described in Part II above, cast a further shadow over the immediate future of the fast breeder and its proposed plutonium fuel cycle. By the autumn of 1977 an International Fuel Cycle Evaluation was underway, to examine different approaches to civil nuclear power that might reduce the problem of controlling the spread of nuclear weapons. Britain was to be an active participant in the evaluation; but the British nuclear people involved were determined that Carter's intervention be stymied, lest it inconvenience their fast breeder plans.

In September 1977 the Select Committee on Science and Technology published the report of its study into so-called 'alternative sources of energy'. AEA chairman Sir John Hill welcomed the committee's recommendation that CFR be built. The following month, at the Royal Institution conference co-sponsored by nuclear proponents and opponents, described in Part II above, Sir Brian Flowers, speaking in the role of a critic in the session on fast breeders, concurred with his co-speaker, the AEA's Tom Marsham, that one large fast breeder was indeed to be recommended. Nevertheless, despite this apparent closing of ranks within the UK nuclear establishment, the government was less and less eager to give CFR the green light. One reason was its preoccupation, from 1977 onwards, with securing some sort of survival for the nuclear plant manufacturers themselves, by immediate orders for units that could be constructed from designs already available. At the time this meant AGRs; neither the competing PWR nor the prospective CFR was at the necessary advanced stage of design work to qualify for the immediate major orders the manufacturers craved.

Added to this was the view expressed by Sir John Hill, that the AEA did not regard the proposed large fast breeder as in any way an experimental plant. On the contrary, it would just be another nuclear power station, of a new design. Behind this confident assertion lay a crucial corollary: if the new plant was just another power station, it would obviously be paid for not by the AEA but by the electricity suppliers, just as they paid for all their other power stations. However understandably appealing this idea was to the AEA, it did nevertheless come up against a basic problem. The CEGB did not want a fast breeder power station – not, at any rate, if it had to pay for it.

Furthermore, the AEA had by this time undermined its own position, by relabelling its proposed plant. It would be not a CFR (Commercial Fast Reactor) but a CDFR (Commercial Demonstration

Fast Reactor). The internal contradiction in this new label did not go unremarked: surely a plant was either commercial or a demonstration plant? The new designation amounted to an admission by the AEA that – *pace* Sir John Hill – the plant would not be in any conventional sense 'commercial'. It would 'demonstrate' the design for a commercial plant; but its electricity output would not be competitive in cost with that from conventional generating plants.

The CEGB let it be known that it would make a site available for a large fast breeder linked to the CEGB system; but it also left no doubt that it had no intention of actually putting up the capital for such a plant. The collapse of electricity demand growth was already embarrassing. The CEGB's excess generating capacity was headache enough as it was, without adding more: especially with the probable aggravation of a novel design. The AEA might get away with pronouncing itself pleased because the PFR's reactor itself was working properly, despite the deep-seated troubles with the steam-generators. The CEGB could not take such consolation. There was no point in being the first kid on the block to run a fast breeder if the reactor could not boil water.

12 Interbreeding

In human affairs it takes two to breed. In nuclear affairs, however, Britain has been a stubborn loner, even when it comes to breeding. British nuclear policy-makers set out independently to develop fast breeders, and thereafter remained essentially aloof from their foreign counterparts, as they did in most other nuclear contexts. There were of course exceptions. In 1970 Britain joined with Federal Germany and the Netherlands in the uranium enrichment consortium URENCO, and in the following year with Federal Germany and France in United Reprocessors. As described earlier, United Reprocessors turned out to be a source more of friction than of collaboration; and the three partners in URENCO likewise each apparently make deals with clients independently of their partners. In fast breeder development, too, this arm's-length relationship between Britain and other nuclear countries was for two decades the norm.

The fast breeder race of the early 1970s was a quintessential demonstration of the competitive atmosphere. Britain, the US and the Soviet Union were first off the mark with fast breeders. The US effort foundered on the embarrassment of the Enrico Fermi fast breeder. In October 1966 it suffered a fuel-melt accident from which it never recovered; the accident inflicted a near-terminal trauma on the US fast breeder programme, delaying it until its economic and diplomatic implications became unmanageable. As a result, despite the teething troubles with the Dounreay fast reactor DFR, Britain could claim with justice in the 1960s that it led the world in fast breeders. Then, however, cumulative technical and managerial problems held up completion of the Prototype Fast Reactor. In the meanwhile the Soviet Union commissioned its BN-350 fast breeder at Shevchenko on the Caspian Sea; it became the first prototype fast breeder to start up, in November 1972. By this time France too had entered the fast breeder stakes. The French Phénix prototype fast breeder started up in August 1973, leaving Britain no better than third in the race. As described earlier, the British fast breeder people started up the Prototype Fast Reactor in March 1974, just before the London conference on fast breeder power stations; but their French rivals forthwith announced that Phénix had attained full power. The fast breeder race was on in earnest; and the OPEC oil price shock, still reverberating after six months, gave the race yet more urgent impetus. The anticipated rush into nuclear power was expected to deplete the planet's uranium reserves within a generation; whoever was first on the market with a commercial fast breeder would win worldwide sales and take the brass ring for global nuclear leadership.

Only twelve months later the picture had changed. The oil jolt helped to trigger an economic recession throughout the industrialized world; soaring fuel prices stunned energy users into a new and thriftier awareness of their previous extravagance. Electricity use stopped increasing; in some countries like Britain it even decreased. Interest rates in double figures made nuclear power, with its staggering capital costs, not more but even less competitive with conventional fuels. The grandiose global vision of an energy future centred on plutonium-fuelled fast breeders began to look less and less plausible.

The fast breeder people themselves held to their faith. In Britain, as described earlier, some AEA staff continued until at least 1977 talking about a vast and immediate programme of fast breeders; others accepted the inevitability of some delay in achieving the plutonium millennium, but remained steadfast in their belief that it would nevertheless arrive, and that Britain would be in the vanguard to embrace it. Others were not so sure. From February to May 1975 the House of Lords Select Committee on the European Communities was taking evidence on a series of papers on 'EEC Energy Policy Strategy', R/3333/74, drawn up by the European Commission. The Lords were far from impressed by the papers, as their report, published the following month, made clear. Under the heading of 'Nuclear Energy', for instance, the Lords commented tartly: 'The Commission's proposals in this field are crucial to its energy programme and the evidence to the Committee leads irresistibly to the conclusion that they are not realistic.'

One witness before the Lords Committee was Walter Marshall, at the time director of Harwell and chief scientist in the Department of Energy. Asked for comment on the EEC policy papers, with their exhortations towards collaborative efforts in the nuclear field, Marshall responded thus:

There are a number of subjects on which by pooling resources they might make some sense. The dominant one in my mind must be the development of nuclear power. There, we are not building the same kind of thermal reactors. Nevertheless, there is some degree of collaboration which might be worthwhile. I think the most important opportunity is on fast reactors. There, I think, a pooling of effort could be helpful. The difficulty is that so many international collaborations are actually difficult to organize and to manage; so you have to balance off the potential advantage with the known difficulties which often arise in international affairs. I personally find it difficult to weigh those two matters together and judge whether there is a net credit or deficit on a project like that.

In due course, force of circumstance was to come to Marshall's aid, and help him to make up his mind. He also added a telling comment, already slightly out of tune with his own colleagues in the AEA: 'I think the first thoughts about the fast reactor were back in the 1950s. If you ask me to make a guess today, I would guess that we will have only two fast reactors operating at the turn of the century at the present rate of progress.' Ere long even this cautious estimate would be revised downwards by 50 per cent.

In the wake of the public furore about the Windscale oxide reprocessing plant, the government gave a commitment that no proposal to build a full-scale 'commercial' fast breeder power station in Britain would go ahead without a major public inquiry. The controversial outcome of the Windscale inquiry and the Parker report did nothing to reassure those who sought a smooth domestic path to the fast breeder. Gradually, very gradually, another possibility began to take shape in their minds.

The Central Electricity Generating Board had already agreed to take a 3 per cent interest in an international group involved in fast breeder development elsewhere in Europe. Called SBK, for Schnell-Bruter-Kernkraftwerk, the group owned 16 per cent of the Super-Phénix 1200-megawatt fast breeder being built at Creys-Malville, in France. SBK was also planning a sister station in Federal Germany, to be designated SNR-2, a full-scale successor to the SNR-300 prototype fast breeder being built at Kalkar, not far from the border with the Netherlands. Given its minuscule holding, the CEGB's active interest in these projects was limited; it was essentially a watching brief, enabling the CEGB to keep an eye on the continental fast breeder activities. But it was also an intriguing precedent, not lost on British fast breeder proponents. As the omens for C(D)FR in Britain grew ever less propitious, the continental link looked ever more enticing.

From 1978 onwards, as described earlier, a further stumbling-block reared up in the path of a full-scale domestic fast breeder in Britain. The official endorsement of the pressurized-water reactor, by Energy Secretary Tony Benn in January 1978 and then, more enthusiastically, by his Conservative successor David Howell in December 1979, was an oblique but daunting challenge to the fast breeder. With electricity demand stagnant and generating capacity in egregious excess the government and the CEGB might be prepared to back the introduction of one new design of reactor; but they were unlikely to back the simultaneous introduction of two different designs. Increasingly it became apparent that the official support for introducing the PWR was also tacitly sidelining the fast breeder.

This is not to suggest any lack of government enthusiasm for the latter. On the contrary: the election of the Conservatives under Margaret Thatcher noticeably revitalized official support for fast breeders – rhetorical support, at any rate. One of Mrs Thatcher's first official visits after her accession to the office of Prime Minister was to Dounreay, on 6 September 1979, to perform the ceremonial start-up of the reconditioned reprocessing line for fast breeder fuel. Asked whether she might soon announce a decision about a commercial demonstration fast reactor she said:

> I have been told: do not just have an inquiry in principle, have it in relation to a specific project. It may be that this would be a faster way of proceeding than having an inquiry in principle. My own personal view is that we should continue with fast reactors, but the government has agreed and is therefore obliged to have an inquiry, and it is not up to me to prejudge the outcome.

It was a warmer endorsement than any given by Tony Benn; but it left plenty of room for domestic political manoeuvres.

Meanwhile, on the international front, fast breeder proponents had been setting aside their differences for more than two years, to repel a common threat. As noted earlier, US President Jimmy Carter had taken a stand against the separation and use of plutonium as a civil fuel, in the light of the possibility that some countries might use such activities to acquire nuclear weapons. Carter's policy statement of 7 April 1977 to this effect had triggered a flurry of top-level diplomacy, heavily influenced by the plutonium lobby in countries like Britain. The outcome was a study called the International Nuclear Fuel Cycle Evaluation or INFCE, in which some sixty countries and five international agencies took part. INFCE was billed as a technical study, to compare different nuclear fuel cycles and their possible use for acquisition of weapons. In the event, however, it was an intense political confrontation between the US and the rest of the world – or at least the nuclear establishments of the rest of the world, with Britain and the AEA well to the fore in defence of plutonium, reprocessing and the fast breeder. When at last the report of INFCE was published, many months later than originally expected, it was a triumph for the plutonium proponents. Proliferation of nuclear weapons, the report declared, was a political problem. It must accordingly be left to the politicians to solve – while the plutonium continued to pour out of reactors and reprocessing plants and fast breeders as its devotees insisted. The international collaboration of the European fast breeder people against the Carter administration may have helped to pave the way for a future collaboration, to defend themselves against the chill wind of nuclear economics and the mounting disillusion of their own governments.

In November 1979 Britain and the US agreed to carry out joint research on certain technical areas, using the Prototype Fast Reactor at Dounreay and experimental facilities at the US government nuclear installation at Idaho Falls, Idaho. Talks were also underway between British fast breeder people and their colleagues in France and Federal Germany. French, German, Belgian, Italian and Dutch fast breeder developments were already being pooled through a company called Serena, which made research data and experience available to all the participants. Britain was exploring the possibility of establishing a parallel company, to be called Fastec – for 'fast reactor technology' – that could collaborate with Serena for further reciprocal assistance internationally.

The key country involved was France. The French Phénix at Marcoule had had its troubles; but it was operating and generating electricity much more reliably than the Dounreay PFR. The full-scale Super-Phénix at Creys-Malville was behind schedule and over budget, but still expected to be on stream by 1984. By the beginning of the 1980s even the British had to concede, with reluctance, that the French were now leading the world – the Western world at any rate – in fast breeders. This was brought home in the most humiliating way possible when France let it be known that Britain would be expected to pay an 'admission fee' of £50 million if it wished to join the European fast breeder link-up. The French demand apparently dampened British enthusiasm for collaboration; but discussions continued – behind the scenes, as usual.

In Britain in 1981–2 the focus of nuclear controversy was the battle over the pressurized-water reactor, and the run-up to the Sizewell B inquiry. The fast breeder people kept their heads down, while the government carried out its own discreet review of fast breeder policy – nothing, of course, as blatant as a planning inquiry, just a quiet reappraisal without any untidy public participation. At length, on 29 November 1982, even as the CEGB and its opponents were gathering themselves for the launch of the Sizewell inquiry, the Secretary of State for Energy, Nigel Lawson, told the House of Commons that:

The Government has now completed its review of the Fast Reactor [initial capitals in the original]. The Fast Reactor is of major strategic significance for the UK's and the world's future energy supplies. It is 50 times as efficient a user of uranium as thermal reactors, such as the Advanced Gas-cooled Reactor and Pressurized Water Reactor, and can create out of the spent fuel and depleted uranium which has so far arisen from our thermal programme fuel equivalent to our economically recoverable coal reserves.

The UK is among the world's leaders in the development of this technology. Through the successful programme of research and development undertaken by the Atomic Energy Authority, which centres on the operation of the Prototype Fast Reactor and associated fuel cycle at Dounreay, we have demonstrated the feasibility and potential of this technology. We have also collaborated with other major countries who have programmes in this field. We are in an excellent position to carry the programme forward and to prepare for the introduction of commercial fast reactors when these are needed to augment our thermal reactor programme.

The Government has therefore decided to continue with a substantial development programme for the fast reactor based on Dounreay and I have asked the Chairman of the Atomic Energy Authority, Sir Peter Hirsch, in consultation with the generating boards, British Nuclear Fuels Ltd and the National Nuclear Corporation to draw up a future development programme which makes the best use of our resources and experience.

So far so good: resounding rhetorical acclaim for Britain's fast breeder effort, indeed a good deal more acclaim than could be justified by the actual track record to date. However, Lawson then went on as follows – reminding some onlookers of the PWR statement by his precursor David Howell three years earlier, with its built-in escape routes:

In common with most other leading fast reactor nations, we now believe that the series ordering phase will begin in the earlier part of the next century, and thus on a longer timescale than we have previously envisaged. We shall therefore have more time in which to develop further the technology and before undertaking the construction of a first full-scale reactor in the UK: and the development programme will be geared to this timescale.

It was a far cry from the lusty assertions by senior AEA staff a decade earlier, foreseeing series ordering of commercial fast breeders by the mid-1980s.

Lawson closed on a revealing note:

The Government and the Atomic Energy Authority have been having exploratory discussions with other countries to establish whether a satisfactory basis for international cooperation can be worked out. The government wishes to see these discussions continue, and has asked the Atomic Energy Authority, in preparing advice about the future programme, to take account of the potential for collaborating with other countries as a means of securing the maximum benefits from this vital development programme.

Not everyone was convinced that the fast breeder programme was so 'vital'. Even the editor of *Nuclear Engineering International*, Richard Masters, by this time had his doubts, as he made clear in a full-page editorial in the February 1983 issue. It minced no words, and summed up briskly the main technical and economic questions undermining the alleged 'maximum benefits' from fast breeders alluded to by Lawson. According to Masters:

The large amounts of money being spent worldwide by the nuclear industry on the development of fast breeder reactors is becoming increasingly difficult to justify. Is this continuing level of expenditure appropriate if one takes a rational view of future trends in energy demand and fuel supply? Will it ever be possible to recoup the vast sums that have been spent and the much greater sums that will need to be spent before the fast reactor can become a commercial option for electricity utilities? . . . Uranium will not be suddenly exhausted or become excessively expensive in the early years of the next century. There will be plenty of time to identify the trend and decide when it is worth ordering FBRs instead of thermal reactors . . .

But perhaps of greater significance to fast reactor economics than the availability of uranium is the fact that with advances in techniques for the storage of irradiated fuel from light water reactors utilities can avoid reprocessing. The uncertain and growing costs of reprocessing are then properly loaded on the fast reactor and with limited reprocessing there will be doubts about the availability of plutonium to fuel a large programme of fast reactors. In these circumstances fast reactors may never be economic.

There still remains the strategic argument but the benefits must still be properly quantified and balanced against the premium it is worth paying for an insurance policy of fast reactors. At present it seems excessive. If the nuclear industry is to win support and acceptance for the fast reactor it will have to provide effective answers nationally and internationally. Evangelical fervour is not a substitute for sound technical argument.

Evangelical fervour nevertheless continued to carry the British government uncritically before it. Ere long Nigel Lawson had been translated to the Treasury, as Chancellor of the Exchequer. His successor as Secretary of State for Energy, Peter Walker, revealed in September 1983 that:

The government has decided to open formal negotiations to seek agreement on joint development of fast reactors with France, Germany, Italy, Belgium and the Netherlands . . . However, we are also conscious that countries outside Europe, particularly the US and Japan, are also experienced in this field. We are therefore keen to keep open the possibility of extending this international collaboration outside Europe when the time is right.

On 10 January 1984, with no further discussion in Parliament or anywhere else in public, Walker and the part-time chairman of the AEA, Sir Peter Hirsch, sat down in Paris and signed an intergovernmental memorandum of understanding with their opposite numbers from France, Federal Germany, Belgium and Italy. The memorandum committed the participants to pool their work on fast breeder research, development and design. It also set the stage for a series of further agreements on joint pursuit of the fast breeder. Three such agreements were signed within the following two months, between the electricity supply organizations, the national nuclear organizations, and the fuel cycle companies of the participating countries. British signatory bodies were the CEGB, the AEA, the National Nuclear Corporation and British Nuclear Fuels.

According to press reports, the agreements foresaw collaboration on the construction of three full-scale 'demonstration' fast breeder power stations, in France, Federal Germany and Britain – probably in that order. The first two would apparently be the plants earlier designated as Super-Phénix 2 and SNR-2. The CEGB was reported to be prepared to take a 15 per cent interest in Super-Phénix 2, in a joint venture with Electricité de France. Where the money would come from, no one said: nor what effect this entire unheralded plunge into an international plutonium nexus would do to the oft-reiterated undertaking by successive British governments to submit any full-scale fast breeder to a public inquiry. The available information about the background to the various agreements amounted to yet another re-run of the shopworn rhetoric about the fast reactor, a 'long term energy resource giving both security and diversity of supply'. Designation of all three of the proposed new stations as 'demonstrations' further emphasized the Alice-in-Wonderland economics of the scheme. It was all too evident why the British government had avoided any prior public discussion of the agreements before committing Britain to them.

While the British nuclear establishment was queueing up to sign on the dotted line, the Comptroller and Auditor General was casting a cold eye at the financial management of the AEA, especially at the fast breeder programme. In February 1984 the C and AG published a terse report entitled 'Development of Nuclear Power', expressing unease about the AEA's financial performance; and in response the House of Commons Committee of Public Accounts looked into the matter. The committee chairman asked AEA chairman Sir Peter Hirsch 'the estimated total cost of development' of the fast breeder. Sir Peter replied: 'We have spent so far about £2400 million in 1982–83 prices. The forward development programme, assuming a certain profit for it, again in 1982–83 prices, is estimated to be £1300 million, the total being £3700 million.' Asked 'What have you got for all this money?', Sir Peter continued:

> The main thing we have got is that we have got the expertise in the UK to go forward to build a CDFR and then have a commercial programme. For that money we shall be, we are, in the position to give the UK the option of having a fast reactor capability for producing electricity. We have done a cost benefit analysis of what the country would get out of it, making certain assumptions. Assuming that commercialization of the fast reactor starts in about 2015 and you have a programme of building fast reactors of 1.25 gigawatts electrical for about 30 years, you can estimate, admittedly on making certain assumptions of uranium price escalation, that you would expect benefits of several billions of pounds compared to the cost you would have to pay if you got the electricity from PWRs . . .

The Public Accounts Committee reported on 19 July 1984, with mild strictures about the AEA's financial targets and programme objectives. It was left to the old warhorses of the Select Committee on Energy to point out the real import of Hirsch's evidence. On the same day the Select Committee published the report of its lengthy investigation into 'Energy Research, Development and Demonstration in the United Kingdom'. The Select Committee was by this time steeped in civil nuclear lore, and long past accepting panoramic promises of nuclear jam tomorrow. Its commentary on the fast reactor programme was incisive:

> The scale of expenditure on this project becomes clearer when expressed in real terms. Since 1955–56 some £2400m (in 1982–83 money values) has been voted for fast reactor R&D, and in the twenty years since 1962–63 real expenditure has remained remarkably steady at between £85m and £120m a year. In evidence to the Committee of Public Accounts on 2 April 1984, the Chairman of the UKAEA estimated that a further 25–30 years and additional R&D expenditure of £1300m (in 1982–83 prices) will be needed to reach the stage 'where one hopes to obtain a commercial station'. To this figure must be added £2 billion construction costs for a commercial demonstration reactor and £300 million for reprocessing facilities, giving total estimated further expenditure of £3.3 billion and a cumulative figure of £5.7 billion. This implies that at present the fast reactor is roughly half-way through a perceived 60-year research, development and demonstration programme. It is interesting in this context to recall that in 1959 the then Parliamentary Secretary to the Minister of Power gave to the House of Commons 'about 1970' as the anticipated date for commercial operation of a fast breeder reactor. As recently as 1976 the UKAEA told the Royal Commission on

Environmental Pollution that it envisaged some 33 GW of fast reactor capacity in place by 2000 . . .

The Select Committee declared itself

all the more concerned to note that for several decades Ministers have been content to rely on advice coming almost exclusively from the UKAEA about the scale of the fast reactor programme. Accordingly, we welcome the Secretary of State's decision to review the programme in depth. We regret, however, that none of the details of this review have been published.

The last two sentences were in boldface. The Select Committee then considered the agreements on joint fast-breeder development.

The proposed scale and phasing of the joint programme raise a number of questions . . . which were not satisfactorily answered in evidence. The intention is to construct three Commercial Demonstration Fast Reactors, one each in France, Germany and the United Kingdom. We find this proposal difficult to understand, except on political grounds, since according to the UKAEA's evidence to the Committee of Public Accounts, there would have been no insoluble technical difficulties attached to the UK proceeding with its own programme (which would have required the construction of one CDFR followed by commercial series ordering). Presumably France and probably also Germany could each have proceeded alone on roughly the same basis. In view of this it is surprising that the joint programme will still require three CDFR's to be built – the same number as would have been the case if each country had pursued its own independent path.

The Select Committee then put the following sentence in boldface: 'There appears to be no obvious rationale for this decision.'

Some would call that the definitive assessment of British nuclear power.

Epilogue: . . . to forgive, incomprehensible

The history of British nuclear power is not a pretty sight. There have, to be sure, been occasional high points; but they are distinguished mainly by comparison with the depths of futility usually plumbed. The innocent bystander, contemplating the wreckage, might be moved to sympathy – were it not that the major perpetrators of this shambles have been paid and are still being paid far more money than most of us will ever see, are sporting CBEs and KBEs and knighthoods, and are received with fresh enthusiasm every time they surface in Whitehall and Westminster touting another sure-fire nuclear winner.

In the coming weeks, months and years Britain's nuclear Marx Brothers will be pressing for official backing for yet more billion-pound nuclear long-shots. They will not settle merely for a PWR power station at Sizewell, but will also have their sights on further PWR stations at Hinkley Point, Winfrith, Druridge Bay in Northumbria, and even – heaven help us – Dungeness. They will be wrestling with the construction of the long-delayed Thermal Oxide Reprocessing Plant at Windscale/Sellafield; they may already be concocting plans for yet another. They will be seeking to establish a variety of disposal sites for the radioactive wastes they have created, while they hide behind the flimsy façade of their own Nuclear Industry Radioactive Waste Executive, NIREX, set up with a staff of seven and a Board of Directors of fifteen.

They will be promoting the construction of a 'commercial' fast breeder reprocessing plant at Dounreay, while pumping British government funds into full-scale fast breeders in France and Federal Germany. The British fast breeder promoters will feel right at home in France: Super-Phénix, the precursor of the projected new full-scale plants, is already three years behind schedule, will cost half as much again as originally anticipated, and produce electricity twice as expensive as that from other French nuclear plants. In due course the British fast breeder promoters will be advancing for the umpteenth time their proposal to build a full-scale fast breeder in Britain. In the meantime they will continue to spent £100 million a year of taxpayers' money in pursuit of this futile chimera.

How long must this surreal charade go on? How long must the British public wait for their government to ask the nuclear power lobby some basic, hard-nosed questions, and insist on credible, realistic answers? No other industry has been so obsessively coddled by its official mentors, backed without stint or hesitation for so long through such a chronicle of arrogant ineptitude. It is time and past time that Britain's nuclear bunglers were finally brought to book.

A note on sources

As its subtitle indicates, *Going Critical* is not an official history of British nuclear power. It is a history of certain key aspects of British nuclear power, told from outside the British nuclear establishment. It does not pretend to be comprehensive; the whole story has yet to be told.

The great majority of the material in this book is drawn from my personal experience as a critical observer of the British nuclear scene from 1969 onwards, assisted by my chronological files of cuttings from the national and international press, which date back to that time and now occupy upwards of two cubic metres of study space. For material relating to events before 1970, I have drawn on four books that must be required reading for any serious student of the British nuclear scene. Margaret Gowing's magnificent official histories, *Britain and Atomic Energy 1939–1945* (Macmillan, 1964) and *Independence and Deterrence* (Macmillan, 1974), set the standard for both scholarship and readability. It must be hoped that she and her invaluable colleague Lorna Arnold can one day tell the entire story to date as it deserves to be told. As it is they are still immersed in the crucial events of the 1950s; all students of nuclear affairs will wish them well in the daunting task that lies before them. *Nuclear Power: Its Development in the United Kingdom* by R. F. Pocock (Unwin Brothers and the Institution of Nuclear Engineers, 1977) was written to celebrate the twentieth anniversary of nuclear power in Britain. As might be expected, it is a view from the inside, from where the picture looks rosier than it does from the outside. It is nevertheless entertaining and useful. *The Nuclear Power Decisions* by Roger Williams (Croom Helm, 1980) is less polite. Williams is unimpressed by the said decisions, and by how they were taken and implemented. His book is a thorough academic analysis of the events and their provenance, exhaustively referenced and annotated; it should be much better known. *Nuclear Power and the Energy Crisis* by Duncan Burn (Macmillan, 1978) is not polite at all. Despite its panoramic title it is in substance an irascible diatribe on behalf of the pressurized-water reactor and against the advanced gas-cooled reactor; but it incorporates a fascinating blow-by-blow account of the infighting in the nuclear corridors of power, provided that one discounts the more far-fetched assertions about the virtues of the PWR. For this earlier material I have also drawn, of course, on the annual reports of the Atomic Energy Authority and the Central Electricity Generating Board, and other official sources as indicated in the text.

Where documents, for instance Select Committee reports, are cited in quotation marks in the text I am quoting from the originals; other citations are paraphrases. Quotations from individuals are drawn from the cuttings files, from the industry trade press, especially the leading British industry monthly *Nuclear Engineering International*, and occasionally from my own notes of meetings, press conferences and other gatherings at which I was present. Unless explicitly attributed to others, inferences, glosses and editorial comment are of course my own views and my own responsibility.

Walter C. Patterson

Index